Weather Whys

Weather Whys

FACTS, MYTHS, AND ODDITIES

Paul Yeager

A PERIGEE BOOK

A PERIGEE BOOK
Published by the Penguin Group
Penguin Group (USA) Inc.
375 Hudson Street, New York, New York 10014, USA
Penguin Group (Canada), 90 Eglinton Avenue East, Suite 700, Toronto, Ontario M4P 2Y3,
Canada (a division of Pearson Penguin Canada Inc.)
Penguin Books Ltd., 80 Strand, London WC2R 0RL, England
Penguin Group Ireland, 25 St. Stephen's Green, Dublin 2, Ireland
(a division of Penguin Books Ltd.)
Penguin Group (Australia), 250 Camberwell Road, Camberwell, Victoria 3124, Australia
(a division of Pearson Australia Group Pty. Ltd.)
Penguin Books India Pvt. Ltd., 11 Community Centre, Panchsheel Park,
New Delhi—110 017, India
Penguin Group (NZ), 67 Apollo Drive, Rosedale, North Shore 0632, New Zealand
(a division of Pearson New Zealand Ltd.)
Penguin Books (South Africa) (Pty.) Ltd., 24 Sturdee Avenue, Rosebank, Johannesburg 2196,
South Africa
Penguin Books Ltd., Registered Offices: 80 Strand, London WC2R 0RL, England

While the author has made every effort to provide accurate telephone numbers and
Internet addresses at the time of publication, neither the publisher nor the author assumes
any responsibility for errors or for changes that occur after publication. Further, the publisher
does not have any control over and does not assume any responsibility for author or
third-party websites or their content.

First edition: March 2010

Library of Congress Cataloging-in-Publication Data

Yeager, Paul.
    Weather whys : facts, myths, and oddities / Paul Yeager.— 1st ed.
      p.   cm.
    "A Perigee book."
    Includes index.
    ISBN 978-0-399-53570-3
    1. Weather—Miscellanea.   2. Weather—Popular works.   I. Title.
    QC981.2.Y43 2010
    551.6—dc22   2009036204

PRINTED IN THE UNITED STATES OF AMERICA

10 9 8 7 6 5 4 3 2 1

Most Perigee books are available at special quantity discounts for bulk purchases for sales
promotions, premiums, fund-raising, or educational use. Special books, or book excerpts, can
also be created to fit specific needs. For details, write: Special Markets, Penguin Group (USA)
Inc., 375 Hudson Street, New York, New York 10014.

Contents

Introduction

THE WEATHER IS THE MOST universal of all topics, and how could it not be? It affects everyone's life every day.

I'm not talking about just the obvious ways—how we all check the weather before we go to sleep or upon awakening to decide what to wear, whether we'll need to allow for more travel time, or whether it will be sunny for the family reunion. Although this type of information is important, it trivializes the true impact of weather.

The weather affects—and sometimes dictates—our moods. Many people I know aren't happy unless the sun is shining and the air is warm, and while I have an appreciation of a sunny, warm day, the blind acceptance that the only nice weather is a sunny, warm day is like saying that the only good music is a happy, optimistic song.

A cloudy, cool, and breezy day can be refreshing and revitalizing, bringing energy back to the body, especially if it's been preceded by a few days of unlimited sunshine and taxing warmth. It's my favorite type of day; thus, it's the name of my weather blog, Cloudy and Cool (cloudyandcool.com). The sound of rain pounding on a roof and dripping onto the concrete sidewalk with an irregular beat can be more soothing than any spa treatment. Add gentle rumbles of distant thunder, and the stage is set for a perfect night's sleep.

The weather is also a living piece of artwork. The sight of undisturbed snow in fields, yards, and tree-lined streets gives us the opportunity to walk through one of nature's postcard moments. In addition, the weather allows us to relive the days of our childhood, lying on our backs and using our imaginations as white clouds change shape and drift past. A product of modern technology, the satellite image has given us the opportunity to see the weather from above, which none of our ancestors could do and early scientists could only dream of doing. We can view the curl of a blizzard, the symmetry of a hurricane, or a lone thunderstorm in an otherwise clear sky.

We're all interested in the weather; it's impossible for us not to be. Some of us just don't realize it.

Don't get me wrong: Some of us know it well—very well. Weather enthusiasts, as they're sometimes referred to, are fascinated by the weather, perhaps chasing tornadoes or obsessing over snow or thunderstorms. Enthusiasts want to know everything there is to learn, including how the atmosphere works, when the next storm is coming, and what's happening in other parts of the country. They watch every relevant show on television, from the Weather Channel to TLC (the Learning Channel) to the Discovery Channel to the History Channel. They turn their obsession into a hobby or profession, and if it seems as if I'm intimately familiar with this group, it's because many people would consider me an enthusiast—although they would most likely refer to it by the less complimentary term: *weather geek*. I'm not easily insulted, though, and even I sometimes affectionately refer to this group as weather geeks.

Many other people are what I call the *weather curious*. This group is overtly interested in the weather, especially their local weather. They wonder how tornadoes form, why it's windy after every snowstorm, how it can snow at 40°F, and why this winter was stormy and last winter was boring. If they see an interesting weather-related show on a Saturday afternoon, they'll watch it. They're not driven enough to turn their curiosity into a full-blown hobby or profession, but if they meet a meteorologist at a cocktail party, they'll spend much of the evening asking questions. Believe me, being the person to whom they're asking the questions, I've met many of them. Of course, being an enthusiast, I'm glad to talk.

The other group interested in the weather is everyone else—even if they don't know it. They might have an intellectual curiosity about how the atmosphere works; however, they most likely don't have enough time to watch television shows or find someone to answer their questions. They might not think about the weather at all, but they do wonder why orange prices are four times higher than they were two weeks ago. They're pleased to learn that it's the cold weather itself that causes the punter of their favorite football team to kick the ball poorly, not simply the fact that kicking a cold-hardened ball hurts the foot. They might want to learn about historic events, including the role the weather played, or they might want to know about the worst natural disasters to affect their ancestors.

In other words, they're living their own lives, which are affected by the weather; therefore, they're interested in the weather. As a person who has not gone through one day in at least the past 35 years (25 professionally) without thinking about the weather or its implications to our lives (I told you I was a weather geek), I've written this book with the understanding that weather interest varies from the enthusiastic to the curious to the unrealized.

Although the book is certainly not a meteorology textbook, some weather explanation is needed in order to appreciate the mysterious world of weather—how moisture and the uneven distribution of heat turns the atmosphere into an unending display of power and beauty that is unchallenged by anything humans have ever created. Explanations are intertwined with entries throughout the book, and the first chapter, "Bite-Size Morsels: Weather Basics," is devoted to weather fundamentals.

All of us—even those who are allegedly uninterested in the weather—have made, or at least have heard, weather statements that are inaccurate, misleading, or incomplete. I know that many of us think that a high pressure system always brings nice weather, a low pressure system makes it windy, and it can be hot enough to cook an egg on the sidewalk. Chapter 2, "It Always Rains in Seattle: Myths and Misconceptions," corrects many of these common beliefs.

The United States experiences a wide variety of weather, from dramatic and dangerous to localized phenomena that are more interesting than dangerous. Chapter 3, "Weird Weather: Unusual and Surprising Phenomena," discusses fires that *cause* thunderstorms even though it's usually the other way around, why lightning can strike out of a blue sky, and 100°F temperature changes in a single day. Unusual or extreme weather for every part of the country, including Alaska and Hawaii, is highlighted.

Weather history can be thought of in two ways: either as weather events that were so dramatic that they should never be forgotten or as weather that affected a historical event in some way. Although some of those events occurred in our lifetimes, others were the defining moments for our parents, grandparents, or great-grandparents. Chapter 4, "From Our Forefathers to Hurricane Katrina: Weather History," takes a look at these weather moments.

Many of our daily activities cannot be separated from the weather, including one of our national interests: sports. Chapter 5, "Field of Dreams: Sports-Related Weather," documents instances in which the weather influenced—or defined—a particular event or game. While we know that weather affects many sporting events, especially football, it's more surprising to learn that the weather occasionally affects games played indoors.

Meteorology has not always been as scientific as it is today [insert your favorite weather forecaster joke here], and we used to depend on observation and recollections from the past to try to understand weather and forecast it. This ancient weather information often comes in the form of weather wives' tales and weather folklore, and although most of us don't depend on maxims to forecast the weather [insert your second favorite weather forecaster joke here], we still hear such things as "If cows lie down, it will rain soon" often enough to wonder if they're true. I take a look at our ancestors' weather insight through the eyes of a modern meteorologist in Chapter 6, "What Does 'Red Sky at Night' Mean? Weather Wives' Tales."

The final two chapters, "Planes, Trains, and Crops" and "The Weather Is Everywhere," take a look at some of the subtle and surprising ways that the weather affects our lives. Whether it be the extra sweetness in an orange because of a light frost, the fact that cold weather *doesn't* make us sick, or the dramatic difference in relative humidity from outside to inside the house, the weather is, in fact, everywhere.

No wonder we're all interested in the weather.

Bite-Size Morsels

WEATHER BASICS

AS YOU MIGHT HAVE GATHERED from the introduction (it's not too late to go back—you're just on page 1 of Chapter 1), this book is *not* about the science of weather; it's about the weather itself and its effects on us.

Although we can appreciate the weather merely for what it is, from the awe of a massive hurricane rising out of the ocean to the simple beauty of a lone puffy, white cloud in an otherwise clear, sapphire blue sky, our appreciation is likely to grow with knowledge. It's analogous to a book on plants. While we might see the beauty in both a desert cactus and a giant California redwood, the appreciation might be heightened by a basic knowledge of how plants grow.

Because many weather concepts and explanations could fill

their own chapters (or books), we'll call these entries bite-size weather morsels—tasty chunks of weather information that will whet your appetite for the remainder of the book.

Air Today, Gone Tomorrow

If I were told that I had about 600 words to explain how weather works in its most basic form, I would summarize two concepts: (1) why air rises and sinks, and (2) temperature and moisture changes of air as it rises and sinks.

Remember the old saying you learned in seventh-grade science class: Hot air rises, and cold air sinks. It's true, and it's all about the molecules.

Hot Air Rises

As molecules become warmer, they increase in speed; dancing around with this newfound momentum, they move a greater distance. This expands their circle of movement, resulting in fewer molecules in a given amount of space (let's call the air in that defined amount of space a parcel) at a given time. The same type of thing happens on a highway when one crowded lane expands into two: the cars move faster, so in any given segment of highway (a highway parcel), the density of cars is less.

As the air molecules in this parcel spread apart, the parcel of air becomes less dense, which means that it physically weighs less. This lighter air is, of course, more buoyant and begins to rise higher into the atmosphere.

Hot air rises.

Cold Air Sinks

With just 600 words, I don't have words to spare, so we'll keep this one short. The opposite happens with cool air. As the molecules slow, more of them accumulate in any given area, increasing the density and weight of the parcel of air, and this heavier air sinks toward the bottom of the atmosphere.

It's just like on our imaginary highway: traffic becomes more dense when the speed of the cars decrease and the two lanes become one.

Cold air sinks.

Rising Air Cools and Becomes More Moist

If a warm parcel of air were to stay warm as it rose, then it would rise until it reached the top of the atmosphere. Fortunately, it doesn't; otherwise, we would have no clouds or precipitation. Air cools as it rises.

Since the atmosphere generally cools with height, part of the cooling of air as it rises is related to the exposure of the parcel of air to its cooler surroundings, just as a steak cools when exposed to a plate cooler than the 500°F grill. The atmosphere is not always cooler aloft, however, so this cannot be counted on to cool the air.

What can be counted on, however, is that the actual process of rising causes the air to cool *and* become more moist. The air pressure always decreases with height (think thin air in Denver), and as the air parcel rises, it moves into this region with reduced pressure, so the air molecules in the parcel spread out. This slows the molecular speed, cooling the parcel. This is called adiabatic cooling, and it occurs at a rate of 5.5°F per 1,000 feet.

The increase in moisture in the air is because the molecular speed of air molecules, which includes water vapor (water in the form of a gas), helps dictate the amount of moisture in the atmosphere. As these molecular speeds slow, a greater amount of water vapor (remember, this is a gas) collects into the form of a solid (ice or water). This is simplified into saying that the atmosphere can "hold" less water when it cools.

While that statement is not literal, it is true that moisture will condense into the air as it rises and cools, which is the basic process of creating precipitation.

Sinking Air Warms and Becomes Drier

It is not surprising that the reverse processes occur when air sinks. As an air parcel sinks, it moves into an area of greater pressure (not only is the air closer to the ground more dense—weighs more—but the entire weight of the atmosphere is felt at the surface, just as the pressure on the bottom book in a stack is greater than the pressure on the book on the top of the stack). The greater pressure forces the molecular speed to increase, thereby increasing the temperature. As the molecular speeds increase, a greater number of solid water molecules (ice or water) turn into a gas, decreasing the amount of moisture in the air.

KEEP THOSE CONCEPTS in mind as you read the book—warm air rises, cool air sinks, moisture condenses into rising air as it cools, and moisture decreases in sinking air as it warms.

Also, remember that I explained the basics of weather in a little more than 600 words.

Air Mass/Source Region

Air is influenced by other factors in addition to whether it's rising or cooling, such as the warmth of the sun above or the cold of the snow under its little molecules. It doesn't take a meteorologist and complicated computer forecast models to understand that air arriving from Nome, Alaska, is much different from air arriving from Havana, Cuba. These differences are often talked about in terms of air masses and source regions.

An air mass is a large body of air (close to the ground) in which the temperature and moisture characteristics are generally uniform. The nature of the air mass is determined by where the air spent enough time (its source region) to acquire its given temperature and moisture characteristics. For example, an air mass locked over northern Canada in January will become very dry and cold.

Air masses have four temperature attributes: arctic, polar, tropical, and equatorial. Air masses have two attributes related to moisture content: continental and maritime. Each air mass is defined by both a temperature and a moisture attribute. Memorizing them isn't important, but understanding that the interactions of air masses help create the weather is.

Battle Lines Are Drawn: Fronts

Weather explanation is often reduced to its most basic form, such as simply saying "the storm was caused by the clashing of cold and warm air" or "a thunderstorm was caused by cool, dry air colliding with hot, humid air." These simplifications have their place, but we may not speak enough in terms of battle lines when talking about weather fronts, which are true lines of weather separation.

A front is the boundary between air masses of significantly different characteristics, which, as we just discussed, were determined by their source regions. While not every front produces dramatic weather, the boundaries between them fall into three categories.

Cold Front

A cold front is the leading edge of a colder (or cooler), drier air mass, and the degree of temperature and moisture change following the front depends on the source region of the air mass. A cold front is typically indicated on a weather map by a solid blue line with blue barbs on it.

Notice that the definition includes the phrase *the leading edge* of a colder, drier air mass. The coldest, driest air will not typically be found immediately behind the cold front but more toward the center of the air mass. For example, when a cold front moves through Pittsburgh, Pennsylvania, the coldest and driest air might just be moving into Minnesota.

Warm Front

Sometimes, the most obvious answer is the correct answer. So if you guessed that a warm front is the leading edge of a warmer (or hotter), more moist air mass, and the amount of temperature and moisture change following the front depends on the source region of the air mass, then give yourself a gold weather vane. A warm front is typically indicated on a weather map by a solid red line with red nubs on it.

Generally, as is the case with the cold air behind a cold front, the warmest, most moist air is probably not going to be found right behind the warm front; however, warm, humid air masses tend to be more uniform than colder, drier air masses, so the peak warmth and humidity might quickly follow the passage of a warm front.

Occluded Front

Sometimes, the most obvious answer is not the correct one because an occluded front is not the leading edge of a more occluded air mass—whatever that might mean.

Typically, a low pressure system will have both a warm front and a cold front associated with it, and the warm front will usually precede the cold front. In an older, more mature storm, there are times in which the cold front catches up to the warm front. When it does, which is not with every storm, this merged front is called an occluded front.

The air mass following an occluded front is likely to have more similar characteristics to the air mass following a cold front rather than that following a warm front, especially because the air mass is drier, so for simplification purposes, many weather forecasters refer to an occluded front as a cold front. When an occluded front is shown, it is typically indicated by a solid purple line with both barbs and nubs on it.

Convection—and Not in My Oven

Anyone who has spent time listening to meteorologists talk about the weather has undoubtedly heard the term *convection*. It's typically used to describe the process that produces thunderstorms; however, that's only part of the story. Thunderstorms are created by the convective process, but convection does not always produce thunderstorms.

Convection is the vertical transport of heat and moisture in the atmosphere. Hot air rises, and cold air sinks (see "Air Today, Gone Tomorrow" earlier in this chapter), and convection is the process that redistributes unbalanced heating by creating air currents. When convection occurs on a small scale (like over your town), it often results in a rising column of air that culminates in a cumulous type of cloud and perhaps a shower or thunderstorm; however, when the atmosphere is too dry, the result is dry convection, such as in Phoenix in June, when the sun continues to shine while convection is occurring.

Convection is one of the atmosphere's processes for attempting to correct an imbalance of heat, which is created by a number of factors. On a global scale, the sun's rays are more directly aimed at some parts of the globe than others during various seasons, creating an imbalance. Think about the difference between Maui and Maine in December. Imbalance is also caused by water warming and cooling more slowly than the land, which influences the air near the water. (See "Oceans Warm and Cool Slowly" later in this chapter.) On a small scale, an example of an imbalance of heat is a blacktop parking lot on a sunny day, which becomes much hotter than a grassy park next to it.

In a convection oven, a fan is used to assist with the process of correcting the imbalance of heat being generated by the heating mechanism. In the atmosphere, it's a natural process. We might not end up with perfectly browned cookies, but we do get all of the weather across the globe.

It's a close call as to which is more important.

Convergence and Divergence

While it's easy enough to think of weather only in terms of our limited perspective on the ground, the weather is created by the interaction of air throughout the entire depth of our atmosphere. One of the most important factors in creating weather is the merging (convergence) and separation (divergence) of air at various levels of the atmosphere.

When air at the ground converges—as it does along a front, for example—it is forced to rise, since it can't collect in one spot (or along one line, as is the case with the front) or sink into the ground below. Rising air is conducive to the production of precipitation (see "Precipitation Formation" later in this chapter). When air is diverging on the ground, as it does at the center of a high pressure system, then the air over that divergent area is forced to sink because air leaving the diverging location needs to be replaced with air from above. Sinking air near the ground is conducive to dry weather.

Those are examples of convergence and divergence at the ground, but air diverges and converges throughout the depth of the atmosphere, creating various effects on the weather.

Hail of a Storm

For those who live in areas where snow and ice are common, frozen precipitation falling from the sky is not unusual, but hail is much different from run-of-the-mill snow, sleet, or freezing rain (see "Precipitation Types" later in this chapter). Chunks of ice, sometimes very large, fall to the ground with great force, not during the winter but during a warm season. They often accumulate on the ground and can take hours to melt. This can cause incredible damage to crops, buildings, and people, making hail one of the most interesting of all weather phenomena.

It's understandable that rain falls out of the sky when the water droplets grow to a weight that makes it impossible for the air to support them, so how are chunks of ice, which are much heavier than drops of rain, supported by the air long enough to become the size of marbles, golf balls, or—heaven forbid—softballs?

Supercooled water—water in a liquid state at temperatures below freezing (which can't happen on the ground but can happen high in the atmosphere)—freezes on tiny particles of dust (called condensation nuclei). These tiny pieces of ice begin to fall, but the updraft (the current of air extending from the ground into the cloud that creates the thunderstorm—the driving force of the thunderstorm) forces the newly formed ice bits (beginning of hail) higher into the cloud, where more supercooled water freezes on the hailstone.

The stone then falls toward the ground until it is forced upward again by the rising air of the updraft. This process repeats itself, with the hailstone rising and falling inside the towering cloud (sometimes 10 miles tall!) like Ping-Pong balls inside a lottery machine until the stone is too heavy for even the updraft to support it, at which time it crashes through the atmosphere and hits the newest car in the parking lot. With that process, of course, the stronger the updraft, the greater the potential size of the hailstones, some of which reach 4.5 inches in diameter.

High Pressure System

A high pressure system is the area where surface pressures are the highest relative to the pressures around it. The air (in the Northern Hemisphere) diverges (moves away) from the center of the high pressure system in a counterclockwise direction. High pressure systems are usually associated with fair weather; however, it's important to note that this is not always the case (see "High Pressure System Always Means Nice Weather" in Chapter 2).

High pressure systems can be found in all layers of the atmosphere, and a high pressure system near the ground is typically represented as a big blue *H* on a weather map.

Hurricane: Monster Rising from the Calm

One of the most awe-inspiring weather facts is that a hurricane, one of the strongest, most-feared storms on the planet, is born out of rather benign meteorological conditions. In that sense, these meteorological monsters rise from the calm of the ocean, growing to swirling storms with a sustained wind of at least 74 miles per hour (mph) and perhaps strengthening to over 200 mph.

While the atmospheric disturbance that begins the process of hurricane formation is typically weak—something that might barely create a weather stir over land—the potential energy present over the bathwater-like ocean is not. The heat and moisture associated with ocean water temperatures of at least 80°F (often in the middle or upper 80s) are like tons of explosives patiently waiting for someone to light a match. A weak disturbance is the ignition for the fuel; when one of these weak disturbances develops a weak, spinning low pressure system, the potential for growth is extreme.

This growth can occur only in an environment in which the atmosphere is relatively calm, though; upper-level wind will destroy the fragile storm, leaving the potential energy untapped. This is in stark contrast to nature's other most dangerous storms, such as blizzards, massive winter rainstorms, and tornadoes, which are driven by strong wind in the upper levels of the atmosphere.

In a Fog

Weather is often associated with human moods—so much so that it's often used as a scene setter in movies. Name a movie when it didn't rain—often a thunderstorm deluge—during a funeral scene, even though, in most cases, the irony of death on a bright, sunny day would be much more striking.

As far as versatile mood setters, fog is hard to beat. A thin, wispy layer of fog in the still of dawn's light signals tranquillity and hope for a new day, and looking from a plane or mountaintop into a valley filled with a milky fog is one of nature's spectacular scenes. In contrast, a heavy morning fog and drizzle (perhaps occurring in the aforementioned valley) muffles ordinary sounds and darkens the morning, and a dense midnight fog can turn even grandma's house into the set of a Halloween thriller. Fog undoubtedly paints a scene.

Fog is often described as a cloud near the ground. In a sense, it is, because fog is a condensation of moisture at the ground as opposed to in the air, but most clouds are composed of ice crystals, whereas most fog is composed of water droplets. Fog generally forms when the atmosphere is nearly calm, so it's unlike most dynamic weather phenomena. The moisture can be the result of a humid air mass, a recent rainfall, moisture from a lake or ocean, or even healthy vegetation. Fog is more likely to form in lower elevations, where the coolest air settles during a long night, than in higher elevations.

Many of you probably balked when I said that fog most often occurs in lower elevations because you've seen plenty of it when driving through higher elevations. Mountain fog is sometimes a low cloud—one that's being driven through as we rise to meet it—or more of a mist created from falling rain and drizzle with little air movement.

Jargon-ogenisis

Jargon is popular in nearly every field because it's fun to create new words and expressions, especially when they sound impressive. We meteorologists are no different, which is why we talk about cyclogenesis, frontogenesis, and bombogenesis.

Genesis means "origin or beginning," so these words simply indicate the process of a developing storm or a developing front. *Cyclogenesis* is used to mean storm development, and

Common Weather Jargon

Weather jargon can be a simple term that represents a compli-cated scientific process, or it can merely be a nickname for a type of storm. Here are a few examples of jargon, along with short definitions:

ALBERTA CLIPPER: A quick-moving winter storm that moves from Alberta, Canada, through the northern Plains of the United States and into the upper Midwest and the Northeast.

BERMUDA HIGH: A stubborn upper-level high pressure system centered near Bermuda; it often brings a heat wave—a tropical heat wave (sing it with me)—to the eastern part of the United States.

EL NIÑO AND LA NIÑA: The largely misunderstood problem children of weather (*el niño* means "the boy" in Spanish, and *la niña* means "the girl"), indicating warmer- (El Niño) or cooler- (La Niña) than-normal ocean water temperatures in the equatorial Pacific. Both affect the weather globally, often dramatically, but their importance is sometimes sensationalized, especially dur-ing weak El Niño and La Niña events. Like most children, they're mischievous but shouldn't be blamed for *everything*.

frontogenesis is used to mean frontal development. *Bombogene-sis* is, in my opinion, an unnecessary term; it's used when fore-casters don't believe that cyclogenesis captures the intensity of a developing storm, so rather than saying "intense cyclogenesis" or "explosive cyclogenesis," many say "bombogenesis" as if the explosive intensification were of the intensity of a bomb exploding.

(The problem is, since *genesis* refers to the formation of some-

MADDEN-JULIAN OSCILLATION: Another Pacific Ocean water-temperature phenomenon that affects the weather over much of the globe. Not as well known or understood as El Niño, it can produce similar effects. Rather than a warming of water temperatures, it's characterized by pulses of energy that move eastward through the Pacific, even moving into the Gulf of Mexico and the Atlantic, adding fuel to storms and influencing tropical storm and hurricane development.

OMEGA BLOCK: An upper-level high pressure system trapped between two upper-level storm systems, resulting in a weather map sporting the Greek letter omega (Ω).

SIBERIAN EXPRESS: A nickname for an Arctic air mass (developed in Siberia) as it heads southward through North America.

TELECONNECTIONS: The method of long-range prediction based on changes in weather patterns in other parts of the world—sort of a "when this happens there, then that happens here" sort of forecasting. Some long-range forecasts (a couple of weeks) and seasonal forecasts (a few months and beyond) are generated this way.

thing, *bombogenesis* should refer to the formation of a bomb, not a storm.)

Jet Stream

An entire book could be devoted to understanding the jet stream; fortunately, a complete understanding isn't necessary for most of us, but a couple of paragraphs of knowledge isn't too much to ask, is it?

The most important thing to know is that the jet stream is the fastest layer of wind in the upper levels of the atmosphere that both creates and directs the weather. It moves storm systems along, helps them form, and gives them energy. The position of the jet stream also determines where the weather will be tranquil, so it's like the great weather director in the sky.

The jet stream is fueled by atmospheric temperature contrasts, and this fastest layer of wind is typically found 20,000 to 30,000 feet above the ground and is fairly narrow (30 to 100 miles across). The jet stream is at its strongest during the winter, and there is typically more than one branch of the jet stream (in each hemisphere), which split and merge, strengthen and weaken, and cause headaches for meteorologists.

Lightning

We meteorologists often look at the weather differently than does the rest of the population. Rather than seeing a storm that's a danger to tens of thousands of people, as a hurricane might be, we see nature's masterpiece—a symmetrical, spinning collection of clouds that taps into the energy of a warm ocean with unimaginable fury and seems to have a personality of its own. Rather than seeing a tornado as a funnel cloud that produces tragic destruction, meteorologists see all of the atmosphere's energy being focused into a swirling bundle that extends from the cloud to the ground with nearly pinpoint precision. If there's one dynamic weather phenomenon that doesn't receive enough respect, though, it's lightning.

The temperature of a lightning bolt is estimated to be as much as 54,000°F (five times the temperature of the sun), and it forms not as a result of a tremendous, ongoing nuclear reaction (like a star) but as the result of an instantaneous release of electric energy accumulated within our own atmosphere—an atmosphere normally as harmless as a bottle of air.

The precise process for the development of a lightning strike is not definitively understood by scientists, but the process of thunderstorm formation leads to the accumulation of a positive charge in the top of a thundercloud and a negative charge on the bottom of the cloud or on the ground (depending on the theory). Air is naturally resistant to electrical discharge, which is something we can be thankful for unless you're envious of Albert Einstein's hair, so it takes a large difference between the positive and the negative charges for them to discharge in the form of a lightning bolt. When it happens, the result is an instant energetic display that's unmatched in nature.

Some people claim to feel electricity in the air before a severe weather outbreak. While many believe the feeling is just a reflection of people knowing that the forecast is for bad weather—they feel what they expect to feel—with that much potential energy awaiting an explosive release, it seems likely that sensitive people do *feel* the electricity.

Low Pressure System

A low pressure system is—no surprise—the area where air surface pressures are the lowest relative to the pressures around it. This is what we typically think of as the center of a storm system, with (in the Northern Hemisphere) air converging inward toward the center of the low pressure system in a counterclockwise direction. Low pressure systems are typically associated with stormy weather; however, it's important to note that this is not always the case.

Low pressure systems can occur in any level of the atmosphere, and a low pressure system at the surface of the ground is typically represented as a big red *L* on a weather map and has warm and cold fronts extending outward from it.

Oceans Warm and Cool Slowly

Water warms and cools much more slowly than does air or the ground. As a result, oceans and lakes remain relatively warm well into the winter and relatively cool well into the summer. Because more than 70% of the Earth's surface is covered by water, water warming and cooling more slowly has a great effect on the weather, either by moderating the extremes of the season or by adding heat, intensity, and moisture to storms.

The influence of water temperature can be seen in many meteorological ways. Cool water can turn a hot day into a comfortable one at a California beach, courtesy of a sea breeze, or it can turn what would be a 70°F spring day into a raw, chilly day with temperatures in the 30s along the Eastern Seaboard.

Warm-water influences can be dramatic, transforming a run-of-the-mill storm into a winter blizzard or producing several feet of snow as cold air moves over a cold lake (see "Incredible Fall Snowstorm" in Chapter 4).

Ometer-itis: Barometers, Hygrometers, and Anemometers

Meteorologists are all children at heart, loving some, or all, aspects of the weather, as if they could relive the snow days of their youths. As with all children, they have their toys, which are given long names to make them sound important.

A barometer is a tool that no meteorologist could live without; it provides a measurement of the air pressure at any given location. While a measurement at a single location has limited usefulness (see "Falling Pressure Always Means Rain Is on the Way" in Chapter 2), a map showing a compilation of pressure readings over a large area (such as a country or a continent) results in the weather maps that we've all seen, which are as important to meteorologists as powdered sugar is to the doughnut maker.

A hygrometer is the instrument that measures the amount of moisture in the air, which is stored in the form of a gas (water vapor). While it's very generous of the air to share its precious space with water vapor, air can be a rude host. Through various atmospheric processes, most notably its rising, the air often turns water vapor into a solid form (snow or rain) and evicts it.

An anemometer is the instrument that measures wind speed and direction, often called a weather vane. If you're talking about the one in the doctor's office, though, it's probably best to call it a weather vein, and if you're talking about the one that certain television meteorologists own, it's probably better named a weather vain.

Precipitation Formation

If you read "Air Today, Gone Tomorrow" earlier in the chapter, then you already know the basic concept relating to how precipitation forms: rising air cools, resulting in condensation of moisture out of the air. When there is enough moisture present in the atmosphere and enough rising air, this condensation process is significant enough to result in the formation of clouds (typically made of ice) and either snowflakes or water droplets (more often snowflakes). When the flakes or water droplets become too heavy to be supported by the air, they fall to the ground.

If the air is too dry, the precipitation evaporates before reaching the ground. This is called virga, which appears either as streaks extending from the bottom of the clouds or as a gray haze that gives clouds a ragged appearance.

That is, of course, a simplified version of events, and because this is a book about weather (not science), it might be best for us to focus on some of the meteorological factors that produce the rising air that, in turn, produces the precipitation. An earlier part of this chapter talked about some of these: the convergence of air caused by a low pressure system, convergence or divergence of air at various levels of the atmosphere, fronts, and convection. We will talk about another factor later in this chapter (see "Upsloping and Downsloping")—a concept that will be worth remembering while reading the rest of the book.

Precipitation Types

Although it's typically colder higher in the atmosphere than it is at the ground (but not always), the temperature doesn't always decrease with altitude in a neat, organized fashion. In other words, different layers of temperatures are found throughout the atmosphere, and especially during the winter months, these different layers of temperatures can alternate between below freezing and above freezing and have a great impact on the type of precipitation.

What follows are the most common types but not all types. Pay particular attention to the descriptions of freezing rain and sleet because many people don't realize there is a difference.

Rain

Rain is, of course, liquid drops of precipitation, and the drops are at least 0.5 millimeters (mm) in diameter. The development of rain often begins high enough in the atmosphere for it to begin as snow, melting as it falls through the warmer layers of the atmosphere. It occurs most frequently in locations where a high percentage of the population has forgotten to bring their umbrellas.

Drizzle

Drizzle is also liquid drops of precipitation; however, despite widespread belief to the contrary, drizzle is not merely rain falling at a light intensity. It is liquid precipitation with drop sizes of less than 0.5 mm in diameter.

In other words, it's drop size, not how heavily the precipitation falls, that determines whether it's rain or drizzle, and it can, in fact, drizzle with greater intensity than rain. The term *heavy drizzle* is not an oxymoron.

Snow

When ice crystals gather in the upper levels of the atmosphere, they become snowflakes. As long as the temperature in the atmosphere is never warm enough to melt the flakes, they will fall to the ground as snowflakes. Snow is sort of like the New York Yankees of the weather world: people tend to either love it or hate it. There's not much mild acceptance.

Sleet

Pellets of ice hitting the ground is called sleet, and it has an adventurous trip from cloud to ground. This type of precipitation is proof of the layers of temperatures in the atmosphere.

The precipitation starts as snow in the upper levels of the atmosphere, but it changes to rain as it passes through an above-freezing layer of air that is thick enough to melt it into rain. (The thicker the warm layer of atmosphere, the greater amount of time the falling precipitation is exposed to above-freezing temperatures and the greater the likelihood that it will melt.) The precipitation, now in the form of rain, then falls through a thick enough cold layer of the atmosphere to freeze it into ice, not snowflakes but chunks of ice, and it lands on the ground with quite a thud.

Sleet pellets can accumulate like snow and can be very slippery to walk on or drive through, but they're solid and heavy, meaning that they're not likely to accumulate on trees and power lines. It's similar to hail (see "Hail of a Storm" earlier in this chapter) in that respect.

Freezing Rain

Freezing rain is the teenage child of the world of precipitation: not well understood *and* likely to cause trouble. Freezing rain is liquid precipitation that freezes upon contact with a surface on or near the ground. It is not ice falling to the ground; rather, it becomes ice when this liquid makes contact with a surface with a below-freezing temperature.

Freezing rain typically takes a similar path to the ground as does sleet, starting as snow and melting as it falls through a warm layer of the atmosphere. The difference—and this is important—is that the layer of cold air near the surface of the ground is not thick enough to allow the precipitation to refreeze. So, instead of hitting the ground as ice, it hits the ground as rain, where temperatures are below freezing, and becomes ice. If you're outside while freezing rain is falling, you will be hit with water, not ice. It will not turn to ice on your warm skin, but if your clothing has a temperature of less than 32°F, it will turn to ice.

This menace accumulates in an ice-rink-like glaze on anything it touches, including trees and power lines, where an accumulation of this freezing water often adds enough weight to bring them down. (For the record, many power lines are not technically brought down by the freezing rain; they're brought down by the tree limbs that are brought down by the freezing rain.)

Hail

Hail is a hard ice formation that falls from thunderstorm types of clouds (see "Hail of a Storm" earlier in this chapter) and is an indication of a dangerous storm. When small, it can innocently bounce around like Mexican jumping beans. When large, it can leave a path of immense destruction.

For some reason, it always seems to be described in one of three ways: the size of a sports-related ball (marbles, golf balls, baseballs, softballs), the size of common food (peas, grapefruits), or official United States coins (dimes, nickels, quarters, half dollars).

Grauple

Grauple is a type of hail, but it's typically softer, smaller, and not destructive. It forms in a similar way to hail, but it's a kinder, gentler version. The associated storm is typically not dangerous; it might not even be accompanied by thunder or lightning.

Relative Humidity/Dew Point

Humidity is the amount of moisture (water vapor) in the air and is measured in two ways: relative humidity and absolute humidity. *Relative humidity*, as the term might imply, is a measurement of the humidity *relative* to the current temperature, and it's measured in percent. For example, if the temperature is 60°F and the relative humidity is 35%, the percentage of moisture in the air is 35% of the amount possible at 60°F. That's what the guy on the evening news always uses.

Absolute humidity, on the other hand, is a measurement of the humidity in a non-comparative way—the actual amount of moisture in the air—which is measured as a dew point temperature. For example, if the dew point is 35°F, it doesn't matter

A 50% Chance of Precipitation

While precipitation is arguably the most important part of the forecast, forecasters do themselves no favors when they forecast a 50% chance of precipitation.

This is the meteorological equivalent of flipping a coin, and believe me, as a meteorologist with nearly 25 years of experience, I've heard variations of that coin-flip joke more often than a career police officer has heard the standard doughnut joke.

A 50% *chance of precipitation* is not a forecast because it gives equal chances of dry weather and wet weather. What good does that do the person at home who wants to know whether he can use one of the items in his prized umbrella collection? The meteorologist is attempting to indicate a better chance of precipitation than on most days, but a good forecaster will make a more definite forecast than this. The job description includes giving a better forecast than the public could get from watching the comedy channel.

what the temperature is: the dew point is 35°F at any temperature. That's what the meteorologist who does the real forecasting uses.

Most meteorologists describe the difference between the two terms by using the analogy of a container of water, but I think it's better to think of it in terms of a container of caffeinated, carbonated cola. Relative humidity would be analogous to the amount of cola compared to the size of the container; for instance, 30 oz (ounces) of soda in a container that can hold a maximum of 100 oz of soda would represent a relative humidity of 30%, since the container would be 30% filled with soda. If those 30 oz of soda were put into a container that held 200 oz, though, the relative humidity would "fall" to 15%, and if the 30 oz of soda were put into a container that held 30 oz, then the relative humidity would "rise" to 100%.

The dew point would be analogous to the absolute amount of soda, and it would not change based on the size of the container in which the soda was placed. If we had 30 oz of soda, then there would be 30 oz of soda whether it was in a container that could hold 100, 200, or 30 oz.

Measuring water vapor in the air causes confusion (see "It's 100°F with 100% Humidity" in Chapter 2). That's understandable. What's less understandable to me is why we're just dumping the soda from container to container when we could be drinking it.

NOTE: **This is an analogy, not a literal example, because air does not hold water in the same way that a container holds liquid even though the difference between the two is often explained in those terms.**

Upsloping and Downsloping

The Earth is not flat (you knew that you'd learn *something* from this book), and because it's not, air is often forced to go either up or down the side of a mountain by the wind. Meteorologists call that *upsloping* and *downsloping*. (We have a lot of fancy names for weather processes, but these are not two of them!)

Because air that is forced to rise cools and becomes more

moist, and air that is forced to sink warms and becomes more dry (see "Air Today, Gone Tomorrow" earlier in this chapter), topographic (geographic features, such as mountains and valleys) variations greatly affect the weather, sometimes subtly and sometimes dramatically. Rain shadows and enhanced mountain precipitation are signs of downsloping and upsloping, respectively.

One side of a mountain might be foggy and cool with drizzle because the air is being forced to rise, while the other side, where the air is headed down, is drier and warmer with sunshine. Although that's fairly innocent—fog and drizzle instead of sunshine on a given day—upsloping and downsloping can vastly affect a region's yearly climate. In Hawaii, where winds are primarily easterly and the terrain is rugged (that's the nature of living on a volcano, I guess), some windward areas (where upsloping occurs) of the Big Island average over 200 inches of rain per year, and parts of the leeward side (where the air downslopes) average 10 to 20 inches of rain per year. It's the difference between a rain forest and a desert.

Wind Direction

Wind direction is always referred to as the direction from which the air is coming. A north wind (or northerly wind) is blowing from north to south, and a south wind (or southerly wind) is blowing from south to north. That's how wind direction will be referred to in all instances in this book.

This might not make sense to those of us who are used to speaking in terms of direction for travel, since we're focused on our destination. We're more interested in the fact that we're heading to the south than that we came from the north. It makes perfect sense from a weather perspective, though, because we're interested in the type of weather that's arriving, not where the weather we had is going next.

It Always Rains in Seattle

MYTHS AND MISCONCEPTIONS

WE MIGHT SAY THINGS SUCH as "The only thing I need to know is whether it's going to rain tomorrow," but five minutes later, most of us are acting like junior meteorologists who, by the way, must have gotten our degrees from the bottom of Cracker Jack boxes. Discussing the weather is common, but we spit out inaccurate statements as quickly as we spit out the hard bits of the unpopped popcorn.

Some of what we repeatedly hear does have a kernel of truth, of course, so our belief in these weather misconceptions is understandable. For instance, Seattle has enough rain during a typical winter for a duck to dream about a weekend in Palm Springs, but saying "It always rains in Seattle" is surprisingly misleading because it rains less in Seattle than in New York City.

We've heard that Chicago is the Windy City so often and for so long that we probably expect to see deep-dish pizza blowing down Maxwell Street; however, Chicago is not one of the 25 windiest cities in the country.

Some common weather statements are more than misleading; they're just plain wrong. After years of repetition, though, it's easy enough to accept them without scrutiny. For example, it has never been—and could never be—100°F with 100% humidity no matter how many times a sports announcer, a news reporter, or your cranky uncle Joe has said it. It's the same with the magic freezing point, 32°F. We've heard (or thought) "It can't snow since it's warmer than 32°F" with enough regularity to think we need to have our thermometer checked when it's snowing at 36°F.

This chapter attempts to correct some of the more widely

Sling Psychrometer

The best-known type of hygrometer (see "Ometer-itis: Barometers, Hygrometers, and Anemometers" in Chapter 1) is called a sling psychrometer, which is a source of embarrassment for all self-respecting meteorologists.

Weather forecasters pride themselves on their highly technical nature, and the image takes a beating when a meteorologist attaches two thermometers to a chain, one of which has a dampened cloth wick over its bulb, and swings the chain round and round as if playing a game for which a 5-year-old would be scolded. But that's a sling psychrometer, which measures the dew point temperature.

held weather myths and misconceptions. Oh, and for the record, I don't even like Cracker Jack.

The Air Is So Heavy (When It's Humid)

Perception is reality in many instances, but that's not the case with our belief that the air is heavy when it's humid. Humid air is, in fact, lighter than dry air.

When the air is humid, our sweat does not evaporate as well as it does when the air is dry, so our sweat runs, our skin remains moist, and we're warm and uncomfortable. Our perception doesn't change the facts, though, and the atomic components of water (oxygen and a double dose of hydrogen) are lighter than the atomic components that make up air (mainly nitrogen and oxygen).

Molecules are like people—only one per location, please—and when the air is humid, some of these water molecules displace air molecules in a given space (usually referred to as a parcel of air). The result is that this parcel of air is lighter, so the air, itself, is less dense when it's humid.

It just feels heaver because we feel warm, clammy, and disgusting.

Chicago Is the Windy City

Chicago, with its broad shoulders nestled alongside Lake Michigan, is a naturally breezy location. This natural breeziness lends itself to the myth that Chicago is the windiest city in the country; in fact, the nickname the Windy City has belonged to Chicago for generations.

Mount Washington, Vermont, has the distinction of being the windiest city in the country, though, with an average wind speed of 35.1 mph. That's actually a myth in its own right: Mount Washington isn't a city and isn't a fair comparison because it's located on top of a 6,000-foot mountain. If Chicago were second to Mount Washington, then it would still count as the Windy City in my book, but it's not. Chicago is not even in the top 25. In fact, the city's average wind speed of 10.3 mph trails many other non-windy locations, such as New York's La Guardia (12.2 mph) and JFK (11.8 mph) airports. However, because Central Park—which of all the reporting sites best represents the city—technically trails Chicago in terms of wind, New York is not as windy as Chicago.

That's good to know. New York City—aka the Big Apple, the City That Never Sleeps, Gotham, the Empire City, the City So Nice They Named It Twice—doesn't need to dethrone Chicago and be called the Windy City, too.

Every Heavy Snowstorm Is a Blizzard

According to the National Weather Service, which determines such things, a blizzard is when snow or blowing snow reduces visibility to under ¼ mile with a sustained wind (or frequent gusts) of 35 mph or greater for a duration of 3 hours. This means that every heavy snowstorm is *not* a blizzard.

Not only is it incorrect to believe any heavy snowstorm is automatically a blizzard, but falling snow isn't even a requirement for a blizzard. While that might seem like Thanksgiving dinner without roast turkey, it's possible. A blizzard can occur merely with wind-whipped snow that's already on the ground as long as the aforementioned wind and visibility conditions are met. Conversely, several feet of the fluffy white menace could accumulate during a snowstorm and not automatically be a blizzard; in fact, if the wind and visibility requirements are not met, then it is not a blizzard and should not be labeled as one.

The proper assignment of the term *blizzard* is made much less frequently than in the past, which is not surprising in our culture of hype. News reporters and many meteorologists (who should know better) often call any heavy snowstorm a blizzard. Now, *you* know better.

Falling Pressure Always Means Rain Is on the Way

We've all seen barometers. Whether it's one with tiny balls suspended in a liquid, a mariner-like dial with a needle that points to words such as *Dry*, *Fair*, *Rain*, and *Run for your life*, or a basic mechanism with a digital readout, home barometers help perpetuate the myth that falling pressure always means that rain is on the way.

The barometer gadgets, of course, are extremely basic forecasting tools, using only the air pressure reading and pressure tendency (whether the air pressure is rising or falling) to make the forecast. In extreme weather events, when the air pressure is either very high or very low, the gadgets work reasonably well because extremely high pressure is typically associated with fair weather and extremely low pressure is typically associated with a rain-bearing storm. That's not too helpful—more like an obvious statement than a forecast.

The fact that we often simplify the falling of a barometer to a forecast of rain is not useful because barometric pressure rises and falls for various reasons, including the daily temperature cycle. Warm air is lighter than cool air (see "Air Today, Gone Tomorrow" in Chapter 1), so during the warmest part of the day (often a sunny day, since those are the warmest), this lighter air results in a lower barometric pressure; the pressure on the barometer falls—toward the word *Rain* on household barometers. This is part of what is called a diurnal (daily) pressure change. This falling barometer reading does not automatically mean that it's going to rain soon; it can happen with no storm in sight—and often does.

Simplifying a weather forecast to *the pressure's falling, so it's going to rain soon* is like simplifying personal accounting to *I got paid today, so I'm going to be rich soon.* Oh, if it were only that easy.

Hawaii Never Has Hurricanes

If you're under the impression that Hawaii never has hurricanes, then you're mostly correct. Of course, being mostly correct about not having a hurricane is like being mostly correct about

Barometer Basics

If you buy a barometer for personal use, then make sure to follow instructions about calibration. All of the barometric pressure readings used by meteorologists have been adjusted to sea level, which is necessary because air pressure decreases with height.

If the instruments weren't adjusted, then it would be difficult to accurately analyze lines of equal pressure. In other words, Denver, with a high elevation, would always have a relatively low pressure. If its readings (and all the others) were not adjusted to sea level, then your favorite meteorologist (or the person who plays one on TV) would not know where to put the red *L*s and blue *H*s.

believing that buses will stop if you step in front of them. It takes only *one*.

Hurricanes typically move from east to west across the tropical Pacific, and water temperatures to the east of Hawaii are slightly cooler than water temperatures across the remainder of the Pacific. The water temperatures are not low enough to diminish existing hurricanes (although the water is cool enough so that existing hurricanes are not likely to strengthen), but they're not warm enough for this region to be a breeding ground for new storms.

More important than the water temperatures, though, is the presence of a semipermanent Tropical Upper Troposphere Trough (TUTT), which is one of the few cool weather acronyms. This TUTT results in a stronger-than-average southwesterly wind in the upper levels of the atmosphere, and this upper-level wind is as favorable to hurricanes as a trip to Hershey, Pennsylvania, is to a tour bus from Weight Watchers. This wind, in combination with the slightly cooler water, means that most storms will weaken before reaching the islands.

The occasional hurricane, though, especially those that form to the south, will ravage the islands. Hurricane Iniki in 1992 was one such hurricane, and this category 4 storm (the strongest in Hawaii history) caused major damage to the island of Kauai and moderate damage in Oahu.

Heat Lightning

Lightning that occurs without audible thunder is often called *heat lightning*. Many believe this to be a special type of lightning— one that occurs without thunder, often with a clear night sky. In fact, when young, I was told (names have been withheld to protect the guilty) that this lightning was not caused by a thunderstorm but by the heat and humidity of the night air. Part of my childhood innocence has been taken from me: There is no such thing as heat lightning.

All lightning is accompanied by ear-rattling thunder; however, the thunder sometimes occurs so far away that it isn't heard. This type of silent lightning is more likely in the Plains, where the large, flat, open land allows distant storms to be seen but not heard.

It makes me wonder about some of the other things I was told.

High Pressure System Always Means Nice Weather

While we may not know everything about the weather, one thing is for certain: A high pressure system always means nice weather. The weatherman uses a nice, friendly *H*, and he uses blue because a blue sky is on the way. Hooray! (That might be where the *H* comes from.)

The presence of a high pressure system does indicate nice, or at least tranquil (since high pressure systems can bring cold weather), weather often enough that the belief is understandable, but it's not always the case. When a high pressure system results in air flow from a water source, such as a lake or an ocean, it can result in damp, dreary weather with drizzle. When a high pressure system results in overly tranquil weather, mean-

Low Pressure Systems and Wind

Most of us mistakenly assume that a low pressure system causes wind, but technically it doesn't. Wind is not caused by low pressure systems or high pressure systems; it's caused by the difference in pressure (pressure gradient) at any given location. The larger this gradient, the stronger the wind.

The strongest gradient will often occur in a region that is between a high pressure system and a low pressure system, so the high pressure system is at least partially responsible for the wind. However, the low pressure system always gets the blame.

It's all about better marketing.

ing it stops all wind, the result is sometimes dense fog and pol-luted air (see "Deadly Fog and Smog" in Chapter 4).

The results of a high pressure system are sometimes pollu-tion, fog, or rain—not the pleasant *blue* sky we always expect.

Hot Enough to Cook an Egg on the Sidewalk

Not being one to *eggsaggerate*, I'm going to have to say that it's *not* hot enough to cook an egg on the sidewalk.

Sidewalks get much hotter than grassy areas because they absorb heat more effectively, and on a scorching day (100°F or higher) under a blazing sun, the temperature of a sidewalk might hit a whopping 145°F or even 150°F. That's certainly hot enough to begin cooking an egg; however, for an egg to solidify (be cooked), the internal temperature needs to reach approxi-mately 155°F.

If a sidewalk chef could manage to get the egg to reach an internal temperature equal to the sun-baked sidewalk tem-perature (145°F or 150°F), then I'd give him credit for a cooked egg. Of course, that's not *eggsactly* my kind of restaurant. The process of putting the moist egg on the hot sidewalk cools the sidewalk just as adding food (with moisture) to a hot pan cools the pan; the cooling process of evaporation begins. Compensa-tion for the loss of heat on the stove is simple—either wait for the pan to return to temperature or do it my way—turn up the heat.

Temperature adjustment, however, cannot be made on a sidewalk. The high temperature was created by the sun above, not the sidewalk below, and increasing the sidewalk temperature again is not possible except by removing the egg and waiting for the sun to gradually rewarm it.

Indian Summer Is Warm Weather in the Fall

The term *Indian summer* is more popular in the fall than apple crisps and leaf blowers; however, the term is often not used correctly. Indian summer is a period of warm weather following the first frost or freeze; it is *not* just any period of warm weather in the fall.

The term is most likely to be accurately used from the northern Plains to the Northeast, where frosts can occur early enough to be followed by a period of warm weather. It's least likely to be used accurately in the Deep South, where warm weather (high temperatures in the 70s and 80s) often persists well into the fall before any frost arrives.

Of course, it's not a term respectful of Native Americans, so perhaps it should be avoided completely.

It Always Rains in Seattle

If you sell umbrellas and dream of moving to Seattle to make your fortune, then it's a good thing that you bought this book before you bought airline tickets. One of the most firmly entrenched weather myths is that it always rains in Seattle.

Seattle averages 37 inches of rain per year. New York City averages just under 50 inches of rain per year. Admit that you had no idea this was the case, and I'll admit that I understand why you've believed this myth for so many years.

Seattle is, in fact, a very rainy location for a significant portion of the year, generally from the middle of October through April. Typically, a relentless train of Pacific storms brings an average of 30 inches of rain during this time period. The "nice" day of many winter weeks is the day when it rains for only a few hours, allowing the sun to peek through the clouds for part of the day.

Seattle residents should be grateful for the winter rain—unless, of course, they prefer a summer of water restrictions and agricultural losses. If the rain (and snow in the mountains) doesn't come during the winter, then it often doesn't come at all. During the summer, when residents in the East are complaining about needing rain because the grass is turning brown after two weeks of dry weather (see "We Really Need Rain, Since the Grass Is So Brown" later in this chapter), Seattle residents often receive less than 0.10 inches of rain for the entire month. In fact, from May through September, while New York averages 26 inches of rain, Seattle averages approximately 7 inches.

New York rain is spread out more evenly throughout the year, so it's perceived much differently than Seattle rain, which seems to come in a nonstop deluge—the type of rain, in fact, of which myths are made.

To be fair, I need to disclose a piece of information that might help explain the myth: Some locations along the Washington coast, to the west of Seattle, receive more than 200 inches of rain per year—about five times as much rain as Seattle and four times as much as New York. This rain also occurs during the same 6-month period as the Seattle rain, so while it obviously rains with great intensity, it doesn't help validate the myth that it always rains.

It Can't Be Humid When It's Cool

I've often heard it said that it can't be humid when it's cool. Strictly in terms of the miserable, stifling summer type of humidity, it's true that it cannot be humid when it's cool, but in terms of *relative* humidity, it can, indeed, be very humid when it's cool.

Relative humidity readings of 100%, although impossible with temperatures in the 80s, 90s, and 100s (see "It's 100°F with 100% Humidity" later in the chapter), are common with temperatures in the 20s, 30s, 40s, and 50s. When precipitation is falling, a 100% relative humidity reading is common, and it's the damp, chilly, and miserable weather that makes your bone marrow shiver and the humidity noticeable.

Annual Rainfall Totals (in Inches) for Major U.S. Cities

Miami	58.53	Chicago	36.27
Atlanta	50.20	Minneapolis	29.41
New York	49.69	Denver	15.81
Houston	47.84	Los Angeles	13.15
Seattle	37.07	Phoenix	8.29

It Can't Snow with Temperatures Above 32°F

Because snow and ice begin to melt at 32°F, the following commonly held belief seems to make sense: It can't snow with temperatures above 32°F.

If the atmosphere were like a brick, consistent from top to bottom and from side to side, then this might make sense, but the atmosphere is nothing like that. The troposphere, the part of the atmosphere in which the weather occurs, is a changing, moving wealth of air stirred by tremendous temperature contrasts. The temperature not only varies from north to south and east to west near the ground, a fact we are aware of (it's usually colder in Nome than in Naples), but also varies from ground level to above us. That has to be true; otherwise, we've wasted a lot of weather balloons over the years.

With the temperatures across the atmosphere so variable, for us to focus on one lonely, little temperature—the one on the ground where we are—is as self-centered as believing that the Earth is the center of the universe. While it does, indeed, need to be 32°F or colder for snow to form, if the snow falls to the ground through a relatively shallow layer of atmosphere with temperatures above freezing, then it doesn't melt along the way. We wouldn't expect an ice cube to melt when dropped from the freezer to the kitchen floor just because the air temperature in the room is 70°F, would we?

Not only can it snow at temperatures above 32°F, but it can snow at temperatures in the 40s and, in extreme circumstances, at 50°F!

It's 100°F with 100% Humidity

If I had a sling psychrometer (see the box earlier in this chapter) for every time I heard someone spout the myth that it's 100°F (or 90°F) with 100% humidity, I'd have more sling psychrometers than I could shake a thermometer at.

For the relative humidity to be 100%, the dew point temperature must be equal to the air temperature (see "Relative Humidity/Dew Point" in Chapter 1), so for the relative humidity to be 100% at 100°F, the dew point temperature must be 100°F. This isn't possible, so the statement would not pass a meteorological lie detector test.

Dew point temperatures in the 70s indicate high humidity, and dew point temperatures in the United States rarely exceed 85°F (and could do so only in the most humid of locations). This means that 85°F with 100% humidity could happen—in a few places on rare occasions. The statement "It's 75°F with 100% humidity" could be said fairly regularly in humid climates, but I don't see that catching on.

It's Always Less Humid After a Thunderstorm

Saying that it's always less humid after a thunderstorm is like saying that it's always less humid in your bathroom after taking a shower.

It is *sometimes* less humid after a thunderstorm because many thunderstorms occur along a cold front, and the air following a cold front is drier and cooler than the air in advance of the front (see "Battle Lines Are Drawn: Fronts" in Chapter 1). In instances when a thunderstorm marks the passage of a cold front, it does, indeed, become less humid after a thunderstorm, but even that statement comes with a caveat: Thunderstorms associated with a cold front often occur well in advance of the front, so the lower humidity might be delayed for several hours. That's probably too fine of a line to draw; regardless, not all thunderstorms occur with cold fronts.

Thunderstorms also often occur along a warm front, and the air following a warm front is warmer and more humid than the air in advance of the front. Not only might a thunderstorm *not* lead to lower humidity, but it can be the precursor to days of nonstop complaints about how it's not the heat but the humidity that's the problem.

Thunderstorms often occur with the absence of any front or with moisture associated from tropical climates. In both of these instances, with no sign of dry air, rain increases the humidity just as taking a shower in a bathroom with no ventilation increases the humidity—and the likelihood of nicks and scrapes due to shaving in a local fog storm.

While it's not always less humid after a thunderstorm, it is always cooler. Evaporation is a cooling process (just as when you sweat), and when rain falls, the air cools. The decrease in temperature might be short-lived, and the higher humidity might negate the effects of the temperature change, but if you want to make a definitive statement about the weather after thunderstorms, then here it is: *It's always cooler after a thunderstorm.*

Sunshine State? More Like the Thunderstorm State

> If the Florida visitor's bureau were staffed with a few more pessimists, instead of the nickname the Sunshine State, Florida would be called the Thunderstorm State.
>
> More thunderstorms occur in Florida than anywhere else in the country, with the number of days with thunderstorms averaging from 70 to nearly 100 per year. The area with the fewest number of days with thunderstorms is the Pacific coast, with fewer than 10 days on average. The tornado-plagued Plains has an average of 30 to 50 thunderstorm days per year; of course, the storms in Tornado Alley are more likely to be severe.

It's Not the Heat; It's the Humidity

Okay, you caught me. The expression "It's not the heat; it's the humidity" isn't a myth as much as an overstatement, but that's close enough for me.

High humidity makes hot weather much more uncomfortable—that's an accurate statement. Evaporation of perspiration is our built-in coolant system, and when the air is humid, our perspiration (civilized people perspire, not sweat!) doesn't evaporate, and we overheat just as a car with no radiator fluid overheats. However, the "it's not the heat; it's the humidity" line is used to extremes.

The combination of heat and humidity can, indeed, make it feel as if the temperature were actually 100°F or even 110°F when the actual temperature is 85°F or 90°F, but believe me, in interior parts of the West, when it's 114°F with a merciless sun pounding down from the cloudless heavens and no hint of a breeze, not many people are saying, "At least it's a dry heat," which easterners think will solve all of their problems.

They're more likely to say, "Can I have a glass of water?"

It's Too Cold to Snow

When extremely cold weather would arrive in the days of my youth, invariably someone would say, "At least it's too cold to snow." Not only was this of no consolation to a snow-obsessed boy, but it's not an accurate statement.

Bitterly cold air develops in land-locked (or ice-locked) places, such as the North Pole, Siberia, and northern Canada, and air masses from these regions are notoriously dry. For that reason, heavy snow is uncommon when it's bitterly cold; an excessively dry air mass is in place. Cold temperatures are not, however, magical show stoppers, I mean snow stoppers, though. When moisture is thrown into these bitterly cold air masses—which can happen fairly easily in places such as the northern Plains, when a storm has access to moisture from the Gulf of Mexico—incredible amounts of snow can fall.

Ironically, if you believe the myth, the extreme cold is part of the reason for the intensity of the snow. Snow falling at lower temperatures is lighter and fluffier than snow falling at temperatures closer to freezing, resulting in a higher snow accumulation with the same amount of precipitation. The colder it is, the lighter and fluffier the snow is, and this type of snow is much more susceptible to blowing and drifting, creating blizzard conditions more easily.

A more accurate statement would be, "It's probably not going to snow because it's so cold, but if it does, look out!"

It's Warmer (or Cooler) Than Normal

Normal weather is a misnomer, and statements such as "It's warmer than normal" and "It's cooler than normal" are more misleading than informative.

By most definitions, the word *normal* means the state that it *should* be, so *normal temperatures* (or *normal weather*) implies that there should be little variation from this so-called normal weather. If the normal high temperature is 55°F (the nice weather guy on Eyewitness News at 10 told me that), then his forecast high temperature of 63°F is not normal; it's much warmer than normal. Something must be horribly wrong. Perhaps the sun has started to throw off a little more heat. Perhaps this whole global warming thing is happening at rapid speed. Maybe we're all doomed.

All right—maybe we won't react quite that dramatically, but the point remains. The term that meteorologists have unfortunately coined as *normal* should instead be referred to as *average*. This average temperature is based on the weather for the past 30 years, so it averages the days when the high was 65°F with the days when it was 60°F, 50°F, and 45°F. In fact, this "normal" 55°F day is most likely the exception to the rule; most days, in fact, will be either noticeably warmer or noticeably cooler.

In many places, especially during the seasons of transition (fall and spring), a great variation of weather is normal. For instance, in nearly every Pennsylvania March that I've lived through, temperatures have ranged from high temperatures in the 70s or lower 80s to high temperatures in the 30s and 40s at some point in the month. I'd be much more surprised (and concerned) if high temperatures were in the 50s every day next March than if they fluctuated dramatically, as they have over the last 40 years.

Call the guy on Eyewitness News and tell him to use the term *average* instead of *normal*. He'd love to hear from you.

Open the Windows During a Tornado

A commonly held myth is that it's a good idea to open the windows during a tornado to keep the house from exploding.

This myth is based on a belief that the tremendous drop in air pressure associated with a tornado (which is true) will create a huge difference between the pressure inside of an airtight house and the low air pressure outside of the house as the tornado moves through. The higher air pressure inside the house will supposedly push the walls of the house outward and collapse the house, and the myth is that opening the windows will balance the air pressure and save the house.

It's unlikely that the air pressure difference would be great enough to exert significant pressure on the walls of a house, and even if it were, it wouldn't matter. Tornadoes often produce wind in excess of 200 mph, so the strong wind and anything blowing around in it, such as a tree or the neighbor's washing machine, is a bigger threat to the stability of a structure than whether the windows are open or closed.

When a tornado approaches, seek shelter in a storm cellar, a basement, or the lowest interior room of a house—don't run around opening the windows.

San Diego Has the Perfect Climate; It's Always Sunny and 70°F

DISCLAIMER: The following entry has not been endorsed by the San Diego Chamber of Commerce.

San Diego is a beautiful city, and the climate is lovely; however, I would like to know why the city has been singled out as having the perfect climate. Statements similar to "San Diego has the perfect climate; it's always sunny and 70°F" have been said so often and with so much certainty that I question my own sanity when I forecast a high of 90°F or rain. That couldn't possibly happen!

San Diego is more temperate than other major metropolitan areas in California—a reflection of the location of the city relative to the Pacific Ocean. It's warmer than generally chilly San Francisco because it's a little farther inland and lacks a bay that funnels wind into the city. In addition, the water to the west of San Diego is not as cool as the water to the west of San Francisco. As a result, the ocean moderates what would otherwise be blazing heat in San Diego but doesn't turn summer into winter, as it does in San Francisco.

In contrast, much of sprawling Los Angeles is located farther inland than San Diego; with less ocean cooling, Los Angeles is more susceptible to intense heat. Los Angeles also has that annoying smog, which can certainly ruin a perfect reputation as quickly as an exterminator truck in front of a sushi restaurant can. (Perhaps this entry will not be the favorite of the L.A. Chamber of Commerce either.)

The Pacific blue also helps keep San Diego pleasant in the winter; because the winter water to the west of San Diego is not as cool as the water farther to the north, the weather is generally milder. It's certainly not as if the ocean ensures the perfect 70°F weather year round, though.

The permanent sunshine part of the myth is even more startling. The cooling breeze from the ocean is typically accompanied by low clouds and fog, especially in the spring and summer

Other 70°F Locations

Because the 70°F high temperature is so widely perceived to be near Nirvana, here are a few other times and places where the average high temperature is close to perfect:

Pittsburgh, Pennsylvania	middle of May
Tampa, Florida	January
Fairbanks, Alaska	late July
Salt Lake City, Utah	early October
Seattle, Washington	middle of June

months. Even magical San Diego has not been able to convince the gods of the onshore flow that it deserves to be immune from the low clouds and fog that are common along the entire West Coast. Finally, while storms are not as frequent in San Diego as in Seattle (see "It Always Rains in Seattle" earlier in this chapter), winter storms don't avoid San Diego.

San Diego has a wonderful climate, but it's no more perfect, in fact, than non-meteorologically perfect climates such as Santa Barbara, Long Beach, and Santa Monica.

The Second Part of a Hurricane Is Always Worse Than the First

The eye of a hurricane is the relatively calm center (sometimes with a clear sky) of the storm, and given the frequency and size of hurricanes, being in the eye of the storm is a rare experience. (You'd never know that, of course, by novels and movies with hurricanes, since all of the main characters manage to find themselves in the eye of the storm.) The eye wall defines the eye of the storm, and this is the part of the storm with the most intense wind in the hurricane, but the perception that the second part of a hurricane is always worse (more intense) than the first is not accurate.

While the circular eye wall, which is composed of bands of thunderstorms, has the most intense wind of the storm, the wind is not consistent throughout the entire eye wall. Some areas have slightly stronger wind than others because thunderstorms have a natural variability; therefore, the second part of the storm, which is the second passage of the eye wall, *might* be stronger than the first, but it's not guaranteed. The second part of the storm might also be more intense if the trajectory of the wind during the second passing of the eye wall is from the water instead of from the land. Wind from the water is stronger than wind from the land because of less friction. Of course, the second part of the storm might be weaker than the first part for the same reasons.

The reason for this myth is purely psychological—and completely understandable. After experiencing increasingly harsh hurricane conditions for hours, followed by a peak with the incredible wind of the eye wall, there has to be some sense of relief when the weather becomes relatively calm in the eye of the storm. To then have the storm immediately return to its peak

Saffir-Simpson Scale

Hurricane strength is based on the Saffir-Simpson scale, which assigns a category based on the maximum sustained wind speed of the hurricane, ranging from category 1 (weakest) to category 5 (strongest). Storms of category 3 or higher are considered major hurricanes.

Category 1	Wind 74–95 mph
Category 2	Wind 96–110 mph
Category 3	Wind 111–130 mph
Category 4	Wind 131–155 mph
Category 5	Wind > 155 mph

conditions, without the gradual buildup this time, has to be an incredible psychological disappointment. The storm was briefly over, and now it's as intense as ever—and this time, having suffered through the first part, you know how intense it's going to be. The wind is also coming from a different direction, which likely increases for a brief time the amount of debris that might have settled in wind-protected locations and might complete the demolition of weakened structures.

Based on those reasons, I'm sure that the second part of the storm always seems worse than the first part, and if you've lived through any storms and want to continue to believe it, then I won't argue with you. You're entitled.

Seek Shelter from Tornadoes Under a Highway Overpass

Several years ago, a cable channel showed video of a group of people, including storm chasers, I believe, who sought shelter from a tornado *under* a highway overpass. The gripping video recorded a tornado (and the reactions of those trapped) passing over the group. They survived the horrendous ordeal, and one myth has been reinforced by the video: Seek shelter from tornadoes under a highway overpass.

Placing yourself under a highway overpass is one of the most dangerous things travelers along a highway can do when faced with an impending hit from a tornado. Just as the speed of a river increases when the river channel becomes narrow, the speed of a tornado's wind will increase when it's forced through the narrow gap between the roadway and the overpass above. The people who survived in the aforementioned video were lucky; however, they'd have faced less harsh conditions had they not been under the highway overpass.

Ideally, of course, the best plan of dealing with a tornado is getting out of the way, but if that's not possible, then seek shelter in a small interior room of a strong building (for instance, a

Tornado Wind Speed

Tornado intensity is quantified not by wind speed but by the amount of damage done—not by design but out of necessity. Accurately measuring the wind speed of a tornado is extremely difficult, if not impossible. First, a tornado is relatively small, so it's rare that it would pass over an existing anemometer. Besides, even if it did, the anemometer would more likely be destroyed and blown into the next county than record an accurate wind measurement.

Given the popularity and increasing technical efficiency of storm chasing and the ability for Doppler radar to estimate wind speed, the ability to accurately record a tornado's wind speed is increasing, but the measurements cannot yet be depended on to gauge intensity. As a result, meteorologists do their best CSI impersonation: They look at the damage done by the storm and then estimate the wind speed necessary to have caused that damage.

The scale is called the Enhanced Fujita Scale (it replaced the Fujita-Pearson Scale in 2007), and it rates tornadoes on a scale from EF0 (weakest) to EF5 (strongest).

home, not a shed). If forced to remain outside, then lie flat on the ground in the lowest possible location, such as a ditch, and protect your head and face as best you can.

Thunder Occurs After Lightning

When thunder and lightning occur nearly simultaneously, we're usually too busy diving under the nearest large piece of furniture to notice how unusual it is to see the lightning and hear the thunder at the same time. When they occur at almost the same time, it's because the lightning strike was very, very close.

Many of us believe that thunder occurs after lightning occurs because that's how our senses perceive it. We see a lightning strike, and then we hear thunder. Sometimes, it's just a second after, and at other times, the thunder follows the lightning by 10 or 20 seconds.

Light travels more quickly than sound, so we see the lightning before we hear the thunder. How far away the lightning strike occurred can then, logically, be estimated by the amount of time between seeing the strike and hearing the thunder. I've heard different estimates for this, and I don't use any of them. Lightning can strike more than 10 miles away from a thunderstorm (see "Lightning Strike out of a Blue Sky" in Chapter 3), so during a lightning storm, whether the most recent lightning strike occurred 200 yards away or 8 miles away makes no difference. The next strike could make that time you forgot to turn the electricity off before replacing the thermostat seem like a pleasant afternoon.

Lightning Destinations

When we think of the word *lightning*, what appears in our mind as quickly as a lightning strike is the image of a bright, raggedy bolt extending from a towering cloud to the ground below—a strike that can turn a stormy night into the bright of day for an instant. This cloud-to-ground bolt is the most dangerous strike of lightning, and it's initiated by a "leader" strike in the clouds and followed by a downward strike, but it's certainly not the only type.

A similar-looking but lesser-known type of lightning strike is called ground-to-cloud lightning, which is initiated from a "leader" strike in the ground, which produces a bolt of lightning that moves upward into the cloud.

Many lightning strikes do not originate from—or strike—the ground, extending from cloud to cloud (cleverly called cloud-to-cloud lightning), often just lighting the base of the cloud without a visible strike. Cloud-to-air lightning is similar, but the ragged bolt is more likely to be visible because its destination is a cloud-less part of the sky.

Tornadoes Are Unpredictable

The difference between not being able to pinpoint the precise location of something likely to happen and having no clue about when or where an event is going to occur is huge. Fortunately, meteorologists forecasting tornadoes fit into the former, but it's the latter belief that lends itself to the misconception that tornadoes are unpredictable.

Tornadoes are not evil, twisting beasts that magically appear out of the clear, blue sky with no warning, in the same random fashion that a leaf falls from a tree. Specific, well-known meteorological conditions need to be met for tornado-producing thunderstorms to form. The challenge in forecasting tornadoes is that these general conditions occur on a much larger scale than the scale of the actual tornado, resulting in a sort of meteorological version of *find the tornado in the very large haystack* game. It might be a long time before the field advances enough for forecasters to be able to pinpoint a precise tornado location days in advance, assuming it ever happens.

Once thunderstorms begin to develop within the area of tornado risk, though, the skill of forecasters to predict which individual thunderstorm is likely to produce a tornado has greatly improved in recent years—and will continue to do so. The key to saving lives is giving people enough time to get out of the way of these increasingly *predictable* storms.

Tornadoes Don't Occur Near a Lake or Mountain

I'd heard that tornadoes don't occur near a lake or mountain so often when I was young that I would wile away the days dreaming of moving to a place called Lake Mountain or Mount Lakeside, where my parents, my brother and sisters, and I could spend our days in the safety of God's bosom. A few years of studying meteorology erased that pipe dream, and it's probably for the best—I'm not the outdoorsy type.

Large lakes, such as the Great Lakes, affect the weather on a

And the Tornado Oscar Goes to . . .

Tornadoes have historically been a mainstay of movies, but to be fair, tornadoes are now taking a second billing to global warming. The tornado often makes an appearance even in these, as it did in *The Day After Tomorrow*, when multiple, swirling monsters tormented downtown Los Angeles. This could never happen, by the way. While it's not impossible for Los Angeles to experience a weak tornado, the climate is not such that strong tornadoes can occur, and global warming is not going to change that in the next millennium or two.

One of the worst movies I've ever had the pleasure to sit through (it was so bad I enjoyed it!) was called *Night of the Twisters*. In real life, two tornado-producing thunderstorms crossing the same area in the same day is extremely rare, but in this movie, tornado after tornado after tornado rolled over the same poor town on this dark and scary night. At one point, a tornado followed the young star as he was driving. While it's certainly not impossible for a tornado to follow the path of a road for a

large scale, large enough at times for the cool water of the lake to stabilize the atmosphere during the spring and early summer months to decrease the danger of tornadoes; however, a typical lake (like the hundreds of thousands in the United States) has no ability to influence whether a tornado forms or to deflect an existing storm away from a town.

It's the same with mountains. Mountains extending thousands of feet into the air never experience tornadoes (not enough heat or moisture), but the average mountain or hill em-

time—if the road corresponds with the direction the tornado is traveling—this tornado followed the car in the same way that a puppy follows its mother. After each turn of this very curvy road, the tornado would reappear in the rearview mirror. It was an instant B-movie classic.

Perhaps multiple tornadoes are not as rare as we might think, at least on Hollywood lots, because the popular movie *Twister* also featured multiple massive tornadoes. Most storm chasers are fortunate to track half a dozen tornadoes over several *years*, which would have been a slow day for the characters in *Twister*. While the special effects were enjoyable, the tornado scenes were riddled with inaccuracies and sensationalism, including twin tornadoes spinning around the storm-chasing vehicle of our daring stars (who were tucked in safely between them) and tornadoes appearing and disappearing, as if they were connected to a light switch.

bedded in the normal terrain does not deter the formation of tornadoes or deflect them from hitting a town.

The Weather Is Always Boring in California

I forecast the weather for California for many years—often enough to hear some version of the "How hard is it to forecast 'sunny and warm' every day?" joke as often as the average basketball player hears "Hi, Stretch."

Hollywood is known for its drama, and this drama spills from the movie sets to the world of weather; however, the drama, intertwined with such prolonged periods of sunny, uneventful weather, results in the understandable assumption that the weather is always boring in California.

As a meteorologist, I've often thought more in terms of California being the Natural Disaster State than a place of boring weather. Mansion-claiming flash floods, levee-shattering river flooding, unimaginable droughts, massive mountain snowstorms, fatal heat waves, town-eating fires, and multiple-accident-inducing fog are all on the bill if you stay long enough. I didn't even mention that little non-weather disaster: the earthquake.

This sounds like a double-feature drama to me.

We Really Need Rain, Since the Grass Is So Brown

Nearly every part of the country has extended stretches of dry weather during the course of the year, especially during the summer months—long enough to result in brown grass. The brown grass *always* results in a chorus of comments about needing rain ("We really need rain, since the grass is so brown"), but water is not always the vital necessity it may appear to be.

It depends on how you define a need of water. If you need the grass to be green and full (perhaps you're a professional croquet player), then you do, indeed, need rain each time the grass is brown. However, if you're more concerned about whether the community has enough water for its needs, it cannot be measured simply by the color of the grass. Grass has a very shallow root system, so it is much more susceptible to a short-term lack of rainfall than trees or other deep-rooted vegetation; however, the ground water or reservoir from which the town draws its water might be filled to capacity.

It might always be *nice* to have rain when the grass is brown

Brown Grass Watch Is in Effect

Meteorological summer (as opposed to astronomical summer) is generally considered to be the time from the start of June through the end of August because those are the three hottest months. Average rainfall varies greatly across the United States in the summer.

CITY	AVERAGE RAINFALL FROM JUNE 1 THROUGH AUGUST 31
Miami	22.96
Atlanta	12.97
New York	12.68
Minneapolis	12.43
Houston	12.36
Chicago	11.89
Denver	5.54
Seattle	3.30
Phoenix	1.39
Los Angeles	0.20

because it's more aesthetically pleasing (and good for other shallow-rooted plants, such as certain crops); however, it might not be the necessity we assume it is. Similarly, the grass will quickly respond to a small amount of rain because of its shallow roots, so it should not be assumed that rain is no longer needed once the grass turns green after an extended period of dry weather. Reservoirs and ground water levels might still be dangerously low even though the grass looks like a manicured golf course.

Weird Weather

UNUSUAL AND SURPRISING PHENOMENA

THE WEATHER ACROSS THE CONTINENTAL United States is more varied and dramatic than in most countries, and with good reason. The United States is large, with a warm ocean to the east (Atlantic), a cool ocean to the west (Pacific), and tropical bathwater to the south (Gulf of Mexico). Mountains soar to over 10,000 feet, and valleys dip below sea level. Some of the coldest air on the planet is a north wind away, and the country's toes are dipping in the heat and humidity of the tropics. With all of that, strange weather should not be a surprise—it should be expected.

As if the weather across the continental United States didn't provide enough of a challenge for meteorologists, President Eisenhower wanted to make our jobs more difficult, so he added Alaska and Hawaii to the weather mix. Thanks a lot, Ike.

This chapter takes a look at some of the stranger weather phenomena, including dangerous storms, dramatic local variations, and unexpected weather.

Aloha, Skiers

Many thoughts go through our minds when Hawaii is mentioned, including the crystal-clear blue water, monster waves for surfing (see "Surf's Up, Dude—Way Up" later in this chapter), warm sunshine balanced by a refreshing trade wind, and passing showers falling out of a tropical sky. Skiing is probably very low on the list, most likely after we wonder why Spam is so popular in the tropical paradise, but aloha, skiers—it does snow in Hawaii.

The two main mountain peaks on the Big Island of Hawaii, Mauna Loa and Mauna Kea, extend more than 13,000 feet into the atmosphere, which is high enough for snow to occur at some point nearly every winter. Snow even occasionally falls on Haleakala Summit, on Maui, which has an elevation of 10,000 feet.

Snowfall can be heavy, in excess of a foot, and the combination of the snow with the naturally stronger wind of the higher elevations occasionally creates blizzard conditions. The snow doesn't typically last long because the air is dry and the southern sun is so strong.

Even in extremely cold storms, though, snow is rarely seen under 7,000 feet, so the only white that beachgoers and surfers see is the sand, the golf balls, and the umbrellas in their drinks.

Backward-Moving Cold Front

Cold fronts (see "Battle Lines Are Drawn: Fronts" in Chapter 1) are not magical blue lines with barbs that systematically march across the globe—even though that's the way they may appear on the weather maps. Cold fronts develop when a colder and generally drier air mass begins to intrude where a warmer and generally more moist air mass has been residing, meaning that they form where the contrast occurs. In the continental United States, cold air generally develops to the north (thanks, Canada), and the weather typically moves from west to east. Therefore, it should be no surprise that most cold fronts arrive either from the north or the northwest (a more direct shot of cold air) or from the west (less direct); however, cold air sometimes arrives from the opposite direction, the east or northeast, with the result being what is considered a *backward-moving cold front*.

The reverse cold front is called a *back-door cold front* in the eastern part of the United States and is fairly common in the spring and early summer. Remember, a cold front is the leading boundary of a cooler air mass, and an easterly wind will force the cooler air from the Atlantic Ocean westward toward land, creating a cold front heading from east to west.

Although the wind can blow from the east or northeast at any time of year, spring and early summer are when these winds, from the Atlantic Ocean, are most likely to produce a dramatic change to cooler weather. Since water is much slower to warm than is the ground (see "Oceans Warm and Cool

Slowly" in Chapter 1), the Atlantic remains chilly through the spring and into the early summer while the temperature trend is increasing over the land of the Northeast and Middle Atlantic regions.

This east or northeast wind can provide a shocking jolt of cold air to residents from eastern New England southward to Virginia and perhaps even North Carolina.

This marine-cooled air often does not extend very far inland—typically not farther to the west than the Appalachian Mountains—meaning that while residents of major cities along the Eastern Seaboard will have high temperatures in the 50s, 40s, or even 30s, much of the rest of the East will enjoy highs in the 70s or 80s. Typically, the weather associated with a back-door cold front includes heavy low clouds and perhaps dense fog. Some rain or drizzle might occur, but precipitation amounts are generally light.

Cold Front Brings Warmer Weather

The term *cold front* says it all: It's the leading edge (front) of colder, or at least cooler, weather. The TV weather presenter even uses blue to show the front on the weather map because blue is a cool color. Occasionally, though, a cold front brings warmer weather.

Two meteorological factors combine to make this warm cold front occur along the south-central coast of California with more regularity and more intensity than in other parts of the country. From spring through summer and into fall, the weather in this coastal region, which includes the city of Santa Barbara, is cool; a westerly wind from the cool-current eastern Pacific results day after day in low clouds and fog followed by a few hours of afternoon sunshine. While high temperatures just inland—sometimes close enough to travel by foot or bike—are in the scorching 90s or 100s, temperatures often struggle to reach 70°F in Santa Barbara. If it weren't for the marine influence, high temperatures in Santa Barbara would be just as hot.

When the wind shifts to the north, though, as it does after the passage of a cold front, the weather world changes in an instant. Instead of a cool wind from the Pacific, a dry, warm wind from the north suddenly develops. The wind is often dry because many cold fronts in this region do not produce rain. The air is warm because the dry, northerly wind interferes with the marine air that was solely responsible for the cool weather in the first place, and the wind is now blowing from warm land, not cool water.

The sudden warming following the front is magnified by the presence of the east-to-west-facing Santa Ynez Mountains, which are located just to the north of the Santa Barbara area. Air warms as it's forced to sink (see "Upsloping and Downsloping" in Chapter 1), so this air, which already represents significant warming in Santa Barbara, is warmed even more. Temperatures sometimes climb well into the 90s or lower 100s after the cold front passes.

Imagine the poor forecaster who has to explain *that* on the morning news: "Clouds will give way to sunshine today. High 68. Tomorrow, a cold front will arrive, so it will be mostly sunny and hot. High 98."

Sundowner Wind

The effect of dramatic warming with a northerly wind in the Santa Barbara area of California does not always occur because of a cold front. Occasionally, when the typical west-to-east sea breeze diminishes at around sunset, the wind shifts to a northerly direction, resulting in a dramatic change in the weather.

The northerly wind races through the mountain passes, picking up speed (often gusting to over 60 mph), and results in an incredible increase in temperature (as described in "Cold Front Brings Warmer Weather" earlier in the chapter), occasionally causing temperatures to jump from the 60s during the day to over 90°F after dark. This is called *sundowner wind*.

Cut-Off Low

We meteorologists certainly aren't people who like to complain, but if we were, then one of our chief complaints would be about what we call *cut-off lows*. Cut-off lows (sometimes referred to as *cut-offs* or *closed lows*) have ruined many perfectly good forecasts through their stubborn lack of cooperation.

A storm doesn't choose to cooperate or not cooperate, of course; however, a cut-off low is one removed (cut off) from the main steering flow, or jet stream. In other words, the jet stream (see Chapter 1) has little influence over this renegade storm, which is embedded in an environment of light, variable wind flow. This erratic, often slow movement gives the storm a fickle, human-like appearance.

Now, if you were a meteorologist with two storms to pick

from, and one was embedded in a 150 mph west-to-east steering flow and the second was sitting in an area with light (under 20 mph) and variable wind flow, which would you prefer to forecast? I, myself, would take the one embedded in the fast flow— it's moving eastward, and it's going to be there soon. You can have the one in the light and variable flow because it might drift eastward tomorrow, or it might drift westward, or it might stay where it is. Cut-off lows sometimes stay in the same general area for 10 to 14 days; meanwhile, the other storm (assuming 150 mph travel for 2 weeks) would have traveled over 50,000 miles.

Fires That Cause Thunderstorms

No one should be surprised to learn that lightning can cause fires. With a 54,000°F bolt of electricity dropping out of the sky, fires are going to happen. What's much more surprising, though, is that the opposite can occur: In rare instances, fires can actually cause thunderstorms to form.

Simply stated, a thunderstorm is caused by the temperature contrast between the warm ground and the cold upper levels of the atmosphere when enough moisture is present. Warm air is more buoyant than cool air (see "Air Today, Gone Tomorrow" in Chapter 1), and moist air is more buoyant than dry air (see "The Air Is So Heavy [When It's Humid]" in Chapter 2), and when the contrast is great enough, the rising motion resulting from the instability produces clouds and a thunderstorm. At times, the temperature contrast needs a boost from other buoyancy-increasing factors, such as a cold front, a warm front, the heat from the sun (which warms the ground), or cool air associated with an upper-level disturbance.

In rare instances, however, heat from a large fire adds enough of a boost to the other atmospheric factors to result in a thunderstorm. In Florida, in 1998, a sea breeze front that was too weak to create a thunderstorm on its own was able to create one over a location where a large fire was raging. Thunderstorms caused by fires happen more frequently in Alaska, where the fires can be large and intense enough to add an incredible amount of heat, and where upper-level temperatures are cool enough to create the needed contrast.

Needless to say, if a firefighter has to deal with a wildfire, this is the kind he would prefer—one that also creates rain to help put the fire out. As rare as this phenomenon is in Alaska, it's even more rare in the western part of the United States, where fires are a serious threat each and every summer. In this region, both the moisture needed for thunderstorms and the cool air aloft are lacking, even when there is heat from a large fire.

Fourth of July Skiing

Holiday ski trips are as popular as diet foods on January 2, with most of the trips focused on the winter holidays (Thanksgiving through Valentine's Day); however, ski season in the western mountains (the Pacific Northwest, Rockies, and Sierra) sometimes begins before Halloween and often lasts through Memorial Day. Believe it or not, even Fourth of July skiing is occasionally possible.

The 2005–2006 winter produced record-high snowfall totals in parts of the California Sierra, and much of the snow fell during a snowy spring. Squaw Valley's website (squaw.com) reported its final snowfall (5 inches) on May 27, bringing the season total to a whopping 628 inches, which is over 50 feet of snow. The depth of the snow (bases were more than 20 feet deep at times) and the lateness of the snow allowed for the ski season to extend to the Fourth of July for at least one resort (Mammoth).

With the melting snow, it must have been difficult to keep the fuses for the fireworks dry.

Ho-Hum, Another 5 Feet of Snow

Snow in the western mountains, particularly the Sierra of California, falls at a rate that is unfathomable to the rest of us—at least those outside of the Great Lakes (see "Incredible Fall Snowstorm" in Chapter 4)—often piling 3, 5, or even 10 feet of snow during a span of just a few days.

Much of the reason is the topography, with monstrous, ragged mountains that are virtually connected to the largest ocean in the world. Every bit of Pacific moisture associated with a storm is turned into ski-loving powder. Another reason is that, unlike most of the other country, storms typically come in bunches. While a snowstorm in the Midwest or East is typically a loner, or at least separated from another by a couple of days, in a stormy weather pattern, storms in the Pacific bring moisture inland with unrelenting fury.

During a storm-filled 5-day span in March 2005, 80 to 100

There's Snow Fence Like a Snow Fence

Many areas of the country use snow fences to help control the amount of snow that accumulates on highways; it's often a wood picket fence, with space between the pickets, several hundred yards away from the roadway. This basic concept works surprisingly well.

The fence is most effective when stationed perpendicular to the predominant wind in the winter; the air slows down as it is forced to go around or through the fence. Reducing the wind speed means that the snow suspended in the air falls to the ground, accumulating to the leeward side of the fence. As long as the fence engineer has left enough room between the fence and the roadway, the snow will accumulate innocently between the fence and the road.

inches of snow fell in the higher elevations of the California Sierra. That's roughly 7 to 9 feet of snow in 5 days! The elevations of these mountains often exceed 7,000 feet, but so do the ski resorts. Not only do some people live there, but others flock there by the thousands to enjoy the benefits of a better snow machine than man could ever create.

Hot, Moist Weather That *Causes* Fires

We all know how hot, dry weather can contribute to fire danger, but there are rare instances when hot, moist weather can actually cause fires to occur.

The combination of moisture, hot weather, and common landscaping mulch can lead to fires through the process of spontaneous combustion. Mulch, which is chopped wood, is an organic product, and the moisture from rain causes the mulch to begin to decompose. Decomposition of organic products produces heat, and this heat in combination with hot weather will occasionally be enough to start a fire. Hot weather and thunderstorms are common in much of the country in the summer, making mulch fires a threat.

A spontaneous mulch fire is much more likely to occur in a large pile of mulch than in a few inches around a home (but it can), and it's also more likely to occur in mulch that's not been disturbed. Therefore, mulch fires are more likely to happen in a manicured tree box in a strip-mall parking lot than in a busy playground.

A similar spontaneous combustion fire can occur in a bale of hay in a sun-drenched hayfield as well, when methane gas (presumably from animal waste) ignites from summer heat.

Hybrid Storm

Storms that are tropical in nature (for example, named storms, such as tropical storms and hurricanes) are so much different from extra-tropical storms (which, ironically, means a non-tropical, standard, run-of-the-mill storm, not one that is more tropical, which the prefix *extra-* might imply) that neither typically thrives in the environment that creates the other. For this reason, the merger of the two types of storms, tropical and

Perfect Trail of Destruction

The so-called perfect storm, which is also called the Halloween storm and the 1991 Unnamed Hurricane, was a remarkable storm even though it was not meteorologically unique.

The storm killed 12 people and caused an estimated $230 billion in property damage in the United States alone—damage was over $300 billion in the United States and Canada combined—as it meandered through the eastern Atlantic for several days from late October through early November. The greatest amount of damage took place from Massachusetts to New Jersey, mainly caused by a strong wind (peak gust of 85 mph in eastern Massachusetts) and monster waves (20 to 30 feet) from North Carolina to Nova Scotia. The waves produced coastal flooding, structural damage, and beach erosion.

extra-tropical, is a fascinating weather phenomenon. It's called a *hybrid storm*.

The hybrid storm is typically the merger of an extra-tropical storm with the remnants of a tropical storm—not an active (named) tropical storm or hurricane. The incredible heat and moisture from the tropics is added to an ordinary storm system (two types of fuel, if you will). The resulting new storm will have some of the characteristics of both a tropical storm and a regular non-tropical storm.

This hybrid storm has more energy and power from the heat and moisture left from what was once a hurricane or tropical storm than it would have had otherwise. In fact, the hybrid storm might have the strength of a hurricane (sustained wind of at least 74 mph), but it might not have the official hurricane designation because it lacks pure tropical characteristics. The National Hurricane Center, which is in charge of such things, could name the storm as a subtropical storm—for example, Subtropical Storm Andrea of May 2007—but it's not required, as it is for a storm with purely tropical characteristics.

The most famous hybrid storm was called the Perfect Storm, which was documented in a book by Sebastian Junger that was

Perfect (Not in My Book) Storm

The first so-called perfect storm occurred in the Atlantic Ocean in October 1991. The impact of this hybrid was further magnified by a high pressure system to the north of the storm. Although it might seem odd to say that a high pressure system played a role in strengthening a storm, because most people already know that low pressure means a storm and high pressure usually brings nice weather (see "High Pressure System Always Means Nice Weather" in Chapter 2)—there's never been a special called *The Dangerous Mega High Pressure System* on the Discovery Channel—it is not uncommon for this to happen (see "Low Pressure Systems and Wind" in Chapter 2).

These three things—a regular storm, the remnants of a hurricane, and a strong high pressure system—merged to create the powerful storm. The label "the perfect storm" was a clear indication that this type of combination was unlike anything else we'd seen before.

It wasn't.

later made into a Hollywood movie. (See "Perfect [Not in My Book] Storm" on previous page.)

> **NOTE:** A hybrid storm is sometimes called a rogue storm, but *hybrid* is a more accurate term because it implies a *combination* of storm factors. *Rogue* means "unpredictable," which is a sensitive word to us meteorologists—and not an accurate description of the storm.

It's Called *Rapid* City for a Reason

While Rapid City, South Dakota, is named after a creek, it could easily have been named because of the rapid temperature changes that regularly take place in the region. On the morning of January 13, 1911, the temperature dropped a whopping 62°F in just 2 hours, between 6:00 and 8:00 a.m. I hope the people who went to work early took a jacket! This still stands as a 2-hour United States temperature change record, and it may not have been as much of a surprise as you might think, since this followed a 55°F change in just 15 minutes that had happened just 2 days earlier.

Extreme temperatures are common in areas that are not influenced by water, which explains why temperatures of more than 100°F are much more common in the northern Plains of the United States and the Plains of Canada than they are in the more tropical climes of Florida. That's right. It's often hotter in North Dakota than in Miami during the summer. (I know, I know, it's a dry heat; see "It's Not the Heat; It's the Humidity" in Chapter 2.)

During the winter, the coldest air in the continent is often just a north wind away from the northern U.S. Plains. Cold air develops in the arctic regions of Canada (and Siberia, from where North America gets some of its bitter cold air) during the winter as quickly as politicians forget about Iowa after the caucuses, and there is little opportunity for this bitter cold to moderate (meaning, get less cold) on its trip over snow-covered ground as it heads southward into the northern Plains. When it turns cold in this part of the country, it turns cold enough to kill.

The mild part of the equation is the result of an air mass from a much different source, along with the local topography. A wind from the Pacific Northwest traveling eastward to the northern Plains will bring with it air that originated over the Pacific Ocean, and air that originates over the relatively mild ocean is drastically warmer than air that originates in the arctic hinterlands of Canada (see "Air Mass/Source Region" in Chapter 1).

In addition, the air is forced to sink as it rides down the eastern slopes of the Rockies—and, in the case of Rapid City, the Black Hills of the western Dakotas—on its trek eastward into the northern Plains, and air that is forced to sink is warmed (see "Air Today, Gone Tomorrow" in Chapter 1). In other words, this mild air is further warmed by the local topography, and having ready access to either type of air mass—extreme cold from Canada or extra-warm Pacific air—results in the potential for extreme, and rapid, temperature changes.

Record Temperature Change

While extreme temperature changes might seem impossible to residents in much of the country (because they are), drastic temperature changes are common in the northern Plains and the northern Rockies. The all-time 24-hour temperature change is an even 100°F in Browning, Montana. From January 23 to January 24, the temperature plummeted from 44°F above zero to 56°F below zero.

Please pass the anti-freeze—and make it snappy.

It's Raining Frogs, Not Cats and Dogs

Reports of animals raining out of the sky have been made for as long as there have been reports of fermented liquids, and even though I have never seen a live creature rain from the sky, this is most likely something that has happened.

The most common reports of creatures raining out of the sky are water creatures, such as frogs, salamanders, and fish. They're lighter than most land animals, such as cats and dogs, and they are typically found in large groups, so these are the most likely types of animals to fall from the sky. If just one unfortunate frog dropped out of the sky, it most likely wouldn't be noticed, but if an entire pack of frogs (I doubt it's called a pack, but I'm a meteorologist, not an amphibian scientist) were to fall out of the sky, people would notice.

An updraft, which is a rapidly rising column of air, can drive a strong thunderstorm, sometimes extending 60,000 feet into the air. We can think of it as God's Hoover, and it wouldn't have difficulty picking up small creatures and depositing them in the rain a few miles down the road.

Creature rain is more likely to happen with a water spout, which is effectively a tornado over water, than with a regular thunderstorm, and the possibility of a storm of that intensity picking up frogs should surprise no one. Tornadoes over land can lift a car, which, I believe, weighs more than your average frog.

Lightning Strike out of a Blue Sky

When a huge, dark cloud is approaching, the wind kicks up and thunder rumbles in the distance, and even though we don't always do it, we all know that we should get inside a building, or at least an automobile, to protect ourselves from a potential lightning strike. While it's easy to understand that potential danger, on rare occasions, there can be a lightning strike out of a blue sky.

It's not that lightning can occur for no reason. This type of lightning still occurs with a thunderstorm; however, lightning can occur as far away as 10 miles outside of a thunderstorm. In most instances, it's cloudy, but it's possible that the sky will be blue 10 miles away from a thunderstorm, making for a lightning strike out of a blue sky.

Monster April and October Snowstorms

While many of us whine whenever snow has the audacity to fall after the end of February because we're ready for spring, residents in the Rocky Mountain states often deal with monstrous snowstorms during the middle of fall and late into spring. In fact, the biggest snowstorms often occur during April and October. The reason is the season—or, more accurately—the changing of the seasons.

Intense snowstorms can, and often do, occur during the heart of the winter in the Rockies; however, the jet stream is also at its quickest during the heart of winter, so these storms often move too quickly to produce incredible snow amounts. During the fall and the spring, storm systems are much more likely to move slowly, so once it starts snowing, it takes a long time to stop. The biggest storms are often cut-off storms (see "Cut-Off Low" earlier in this chapter), which might remain in the same general region for days on end, pulling moisture from the Gulf of Mexico westward into the Rockies and producing snow by the feet, especially on the east-facing mountain slopes.

One such storm late in April (21–24) 1999 produced more than 4 feet of snow (52.7 inches) in Lander, Wyoming—the biggest snowstorm on record for that snowy Wyoming town. Happy

Snowy Octobers and Aprils in Denver History

October 1969	31.2 inches	April 1933	33.8 inches
October 1906	22.7 inches	April 1885	32.0 inches
October 1997	22.1 inches	April 1945	28.2 inches
October 1923	17.9 inches	April 1957	25.5 inches

spring! In October 1969, just over 31 inches of snow fell in Denver. During that season, another 20.5 inches fell in the city in March, after only about 15 inches had fallen from November through February.

That type of winter is not going to make too many people happy. Those who love snow want it all winter long, not just in the fall and spring, and those who don't enjoy snow can usually put up with it in the winter but don't want to think about it in the other months.

Snow-Eating Wind

Snow is typically either loved with a passion unmatched in the weather world or despised as much as if it were a white gift from the depths of evil. Both snow lovers and snow haters monitor the same aspect of snow—how quickly it disappears. It might be surprising to learn that disappearing snow doesn't always melt, and its non-melting disappearing act is often assisted by a *snow-eating wind*.

We all know what happened to Frosty the Snowman when he made the unfortunate choice to enter a greenhouse; he melted into a puddle of Christmas sadness. He could have faced a similar fate had he run into a dry wind, but it would not have made for good television. Instead of turning into a puddle, he would have slowly vanished into thin air. This process—when any solid substance turns directly into a gas without turning into a liquid first—is called *sublimation*.

The best known example of a snow-eating wind is the Chinook, or foehn wind, in the leeward (downwind) side of the Rockies of North America (United States and Canada). As air is forced down the side of the Rockies, the air warms and becomes more dry, characteristics not favorable for snow's long-term health. With a depletion of the snow cover and none of the slush and puddles that, say, Buffalo, New York, has to deal with, it appears as if the wind were eating the snow.

Santa Ana Wind

We all know what the Santa Ana is: It's a strong wind in California that causes fires, right? Well, that's close—it *is* a strong wind in California, but contrary to the way that it's typically portrayed in newscasts, the wind does *not* cause fires.

Fires are much more likely to occur while a Santa Ana wind, an easterly wind in the Los Angeles Basin, is occurring, but the wind cannot cause a fire (unless it could rub sticks together like a Cub Scout). The wind speeds are magnified as the wind is forced down the side of the mountains and through canyons and passes, and because the air is coming from the desert, a naturally dry location, it's already dry. The air is further dried by its descent (see "Air Today, Gone Tomorrow" in Chapter 1), so it's extremely dry, sometimes with a relative humidity of under 10%, by the time it reaches the L.A. Basin.

The extreme dryness of the air combined with the dryness of the vegetation after a long, hot summer results in tinder-like conditions. Any spark will result in a fire, which will be driven by the erratic, dry wind; as long as the wind blows, the fire will be difficult (if not impossible) to contain. The result is often that reporters talk about the Santa Ana wind as if it caused the fires.

Everyone Knows It's Windy: Wind Names

CHINOOK WIND: A warming wind that occurs in the leeward (downwind) side of the Rocky Mountains of North America (see "Snow-Eating Wind" earlier in this chapter).

DIABLO WIND: Similar to the Santa Ana wind but for northern California; this wind occurs in the canyons of the East Bay Hills and can cause the same problems for fires that the Santa Ana causes.

KONA WIND: A south to southwesterly wind accompanied by a storm system over or to the west of the Hawaiian Islands; often results in heavy rain, with the heaviest rain often occurring in the normally dry parts of the islands.

NORTHER (SOMETIMES CALLED A BLUE NORTHER): A cold, strong northerly wind in the southern Plains of the United States, especially in Texas, which results in a dramatic change to colder weather. It's often the leading edge of an Arctic air mass from Canada.

SUNDOWNER: A gusty north wind that occasionally occurs along the south-central coast of California, resulting in dramatic warming (see "Sundowner Wind" earlier in this chapter).

KNIK WIND: A strong southeasterly wind that occurs in southern Alaska, mainly in the winter. It occurs near Palmer, which is not far from Wasilla. Perhaps it should be renamed the Palin wind.

This type of wind and fire combination can occur in many regions in the western United States, especially California, but it's called a Santa Ana only if it occurs in the Los Angeles area with a wind out of the east or northeast.

Surf's Up, Dude—Way Up

Two things are certain: Where ocean and sand meet, there will be surfers, and surfers in the western part of the country and Hawaii will be happier than those in the East.

Before explaining the factors that produce the larger waves, it's important to note that there is a difference between *wind waves* and *swell*. The swell is the large, rolling motion of the ocean that can make the shrimp alfredo on the U.S.S. *Vacation* seem as if it had been a very bad idea (please pass the Dramamine). The swell travels the ocean blue, sometimes for a thousand miles. Surfers are interested in the swell, which becomes a large-scale breaking wave as it approaches the coast.

Wind waves, on the other hand, are a reflection of local wind. These are the waves that appear on top of the ocean, so they appear on top of the rolling swell. Wind waves are typically small and choppy but can become large and disruptive during storms. Wind waves are found on inland lakes, where there is no swell. On a windless day at the beach, there may not be any wind waves, but there could still be a large swell. For this discussion, I'll use the words *waves*, *surf*, and *swell* to talk about the swell.

The size of an ocean wave is generally a function of the

strength of the wind, typically created by a storm, and the distance of uninterrupted ocean over which the wind blows. In that sense, all oceans can create big waves. Since Pacific storms head toward the West Coast and Hawaii over a long, uninterrupted fetch of ocean, meaning that there are no land masses of consequence between the wave source and the land, the waves for the West Coast and Hawaii are large. Along the Atlantic Coast of the United States, the wind and storms generally move away from land, so the largest waves are typically directed eastward into the Atlantic, away from the beaches.

Remember, it's the energy from distant storms that results in the large swells; it's not a local storm that produces the swell. The swell can travel many, many miles, so even though the bulk of the powerful, wave-generating Pacific storminess remains to the north and west of Hawaii, the high surf makes it there. In fact, it's easier to have large waves in Hawaii than along the West Coast.

Hawaii is a series of volcanic islands poking out of a deep ocean, as opposed to the West Coast of the United States, which is a large continent with a relatively gentle continental shelf leading to it. The shape of the continental shelf—or the lack of a shelf altogether—is an important factor in wave size. The steeper the shelf, the bigger the breaking swell along the coast. The West Coast has a steeper continental shelf than the East Coast, which serves to maximize the waves—waves, as we just discussed, with generally greater potential. Hawaii, with no continental shelf, leads all three in the potential for waves.

Waves often reach heights of greater than 25 feet on the

north- and west-facing beaches of Hawaii (waves can exceed 40 feet). Waves of 25 feet can also occur along the West Coast, but this is typically limited to times during a monster rainstorm—perhaps one with hurricane-force winds. In Hawaii, it often happens with sunshine and temperatures in the 80s, which is a little more conducive to chasing the big one.

Enjoy the waves; I'll be enjoying an umbrella drink while I look for my sunscreen.

Weather Oxymoron—Desert Flooding

While the Desert Southwest is best known for its heat—dry, ground-cracking, unrelenting, and potentially deadly heat (and the annoying perception that since it's a dry heat, it doesn't feel hot; see "It's Not the Heat; It's the Humidity" in Chapter 2)—flash flooding is often a serious concern. While that might seem ironic (or oxymoronic), it's just part of the natural cycle of the weather.

This region doesn't get much rain; otherwise, it would not be a desert. The average yearly rainfall for Phoenix, for example, is a scant 9 inches. When thunderstorms arrive, though, as they usually do during late summer through fall, the result is heavy rain falling on sun-baked ground—ground that's practically as hard as concrete and lacking in vegetation, which is the type of ground that water rolls off of, not the type that absorbs it.

Desert Southwest thunderstorms usually move slowly, dumping rain over the hardened ground for a prolonged period, making the region a prime location for the threat of serious flash flooding. A partly sunny and hot day can quickly be replaced by flooding, as dry river and creek banks become raging torrents of destructive water.

Winter Heat Waves

Winter heat waves are more common than might be expected, and I'm not talking about the heat *inside* of grandma's house (see "Can You Turn Down the Heat, Grandma?" in Chapter 8). The southern part of the country, while typically fairly comfortable during the winter months, is far enough to the south for legitimately hot weather to occur during the winter. The hot spells are often aberrations in otherwise cool weather, creating a stark contrast.

December 1971 was one of those years in Los Angeles. Nearly every day for 2 weeks, high temperatures were in the 50s in Los Angeles (I know, I know—I can hear the fake cries of sympathy from my northern readers), a time when the average high temperature is close to 70°F, but hot days reminiscent of summer returned in a flash. Record high temperatures were set on December 17–19, and when the temperature downtown reached a high of 95°F on the 18th, the thermometer in some of the valley locations undoubtedly touched 100°F.

Even though high temperatures did not return to the chilly 50s, much more normal weather returned within a few days.

Winter "Hurricanes" in the Pacific Northwest

As you know, I'm a weather geek sometimes, and the proof is that I enjoy watching TLC's *Deadliest Catch*, a reality show set in the Bering Sea in the winter. I don't watch this reality show because I like boats or because I like the drama of clashing personalities or because I like to see huge amounts of crab ripped from the ocean floor. I watch it because of the weather and the waves—some of the most intense on the planet.

The storms are in no way true hurricanes even though the sustained wind often exceeds 74 mph, which is hurricane force; the environment in which they form is the opposite of the breeding grounds of a hurricane. Winter storms in the Bering Sea are produced by cold air and a strong wind in the upper levels of the atmosphere; hurricanes are produced by warm water and little or no wind aloft (see "Hurricane: Monster Rising from the Calm" in Chapter 1). Drop a hurricane into the Bering Sea in December, and it would fizzle more quickly than a firecracker dropped into a swimming pool. The reverse holds true as well; drop one of these Bering Sea monster storms into the middle of a calm ocean in September, and it would wilt as quickly as a bouquet of roses in a sauna.

The storms are monsters, though, producing tremendous rain, snow, wind, waves, and swell, often plowing into the west coast of Canada and the Pacific Northwest. Wind gusts have been known to exceed 100 mph, and waves, especially over the open ocean, can exceed 30 feet—and the storms often follow one after another like cars on a freight train.

Men try to fish in these conditions—and I thought meteorology was a tough job.

Winter Tornadoes

Tornadoes are more common in the spring than in any other time of the year; in fact, meteorologically speaking, spring is often referred to as severe weather season for this reason. Some of the same weather factors responsible for the nasty spring storms return in the late fall and early winter, resulting in a second severe weather season.

Ingredients needed for tornado-producing thunderstorms include high humidity, the contrast between warm air at the ground and cold air above the ground, and upper-level disturbances that produce atmospheric spin (commonly called wind shear, which is a change in wind direction and speed with height). These factors are most common in spring, when humidity and heat have begun to increase in the southern tier of the country, but the chilly upper-level storm systems from the previous winter have not completely finished coming through for the season.

By the latter part of summer, tornado-producing thunderstorms are much less common. While humidity and warmth at ground level remain abundant, the air in the upper levels of the atmosphere is not as cold as in the spring, so the contrast between air at the ground and air above the ground is not as great. Shear-creating upper-level storm systems are also less frequent.

In the fall, though, things begin to change once again—in a way that is opposite to the spring changes. In the spring, it's an increase in summer-like heat and humidity that contrasts with the remnants of a winter storm track that results in the violent weather contrasts. In the fall, it's an increase in a winter-like storm system with the remnants of summer heat and humidity that sparks severe weather. These changes are not typically as dramatic in the fall as they are in the spring, but from November through early December, there is often a second, less intense severe weather season across the United States.

Of course, it's only less intense if it doesn't affect your home or family.

From Our Forefathers to Hurricane Katrina

WEATHER HISTORY

HISTORIC WEATHER FALLS INTO TWO categories—weather phenomena remembered for generations because of their extreme (and historic) nature, and historic political or social events marked, or influenced, by the weather. This chapter takes a look at both.

For events before the early 20th century, weather records are not detailed enough for the type of analysis to which we, in our modern world, have become accustomed. We can, however, through historical accounts, including occasional weather records, often put those events in weather perspective.

For the more recent extreme events—dramatic enough to seal their place in history immediately—we have detailed weather records. Either way, the weather has played a great part in the history of the United States.

Billion-Dollar Freeze

California is typically considered to be a winter destination—a place to go to avoid the bitter cold; we certainly don't think of California as a place with extreme cold.

Generally, weather travels from west to east, so the mild Pacific Ocean usually prevents extreme cold in California, but when a direct north wind develops (north, not northwest, since a northwest wind would have trajectory over the water, tempering the cold), air directly out of western Canada takes a trip southward. In other words, some of the potentially coldest air in the continent decides to take a winter trip toward Disneyland, and its arrival marks a potentially devastating effect on winter crops for the entire nation. A recent extreme example is a freeze in January 2007, which sent temperatures in the Central Valley of California, the location where most of the country's citrus crops are grown, into the teens and lower 20s. Damage to the citrus crop alone was estimated at close to a billion dollars.

Challenger Disaster

Any time an arctic cold front (a cold front that is the leading edge of a bitterly cold air mass) moves through Florida, it's newsworthy because of the likelihood of crop damage. An arctic cold front on January 27, 1986, is one that we'll be talking about—at least indirectly—for generations to come, since cold weather has been identified as a contributor to the disaster of the *Challenger* shuttle on January 28, 1986.

Because of the extreme cold at the time of the launch (28°F in Cape Canaveral, Florida), conjecture about whether the cold might have been a contributing factor was immediate in the meteorological world. The Rogers Commission later confirmed those suspicions; O-rings (circular gaskets), due to faulty design, failed in the cold weather, resulting in the shuttle disaster that killed seven astronauts, including teacher Christa McAuliffe. Ms. McAuliffe would have been the first teacher in space.

Chicago Heat Wave

Heat waves are often slow to develop and slow to dissipate (except in the West, where a slight shift in wind direction often results in incredible temperature differences in a short period of time), resulting in a week or two of unrelenting heat and humidity along with deteriorating air quality, eventually wearing down power grids and, sadly, the weak and the elderly. Heat waves are not typically quick to begin or quick to end, but the intense Chicago heat wave of 1995 was fairly short-lived.

July 12–16 was preceded and followed by relatively nondescript summer weather, but those five days in between were intensely hot and humid, resulting in one of the worst heat waves in U.S. history. Heat-related deaths in the city were estimated to be between 400 and 800, signifying the intensity of the heat. High temperatures peaked from the 12th to the 14th, with three consecutive record high temperatures. It was 105°F at Midway Airport and 104°F at O'Hare on July 13.

While the heat was unbearable, it was the combination of heat, humidity, and an urban heat island that made the week so tragic. Heat indices—how hot it feels when humidity and temperature are considered—exceeded 125°F. The heavily urbanized parts of the city, full of concrete and blacktop, did not cool as much as the more open areas, where the airports are located. With low temperatures in the middle 80s at the airport, low temperatures in parts of the city never dropped below 90°F. In much of the country, a high of 90°F is a hot day; lows of 90°F are largely unheard of outside of the low deserts of California and Arizona.

Unrelenting exposure to the extreme heat and humidity put the elderly, very young, and ill at extreme risk, and this risk was compounded by air-conditioning failures related to power demands as well as a complete lack of air-conditioning in some of the older buildings. Because heat rises, the temperatures in the higher floors of those older buildings must have been over 100°F at night, with daytime temperatures over 135°F.

It's impossible to estimate how many people would have died under these conditions had the heat lasted for 2 weeks rather than 5 days, but that's the only thing that the residents of Chicago had to be thankful for in July 1995.

Deadly Fog and Smog

The valleys in the eastern part of the country often fill with fog during the lengthening nights of fall. In fact, fall is the most common time of year for fog, and the stagnant weather pattern resulting in this trapped moisture near the ground (which is what fog is) can also trap atmospheric pollutants. Before we had a complete understanding of the danger of pollutants and stagnant air masses, this combination resulted in a killer fog/smog in Donora, Pennsylvania.

After a weak cold front moved through the small western Pennsylvania town (located about 25 miles south of Pittsburgh) on October 24, 1948, a high pressure system took control of the weather from October 25 through October 30, resulting in several days of stagnant weather. The fog, which was mixed with the smoke and pollution from the local steel-related mills, resulted in an all-day thick, hazy fog. This was no ordinary fog/smog, though; as the days progressed, the fog and smog became thicker, and residents became ill by the thousands. Some of them, 20 in total, died as a result of the poor air quality.

A lack of understanding of how pollutants were trapped by stagnant weather and a resistance to shutting mills—the livelihood of nearly the entire town—resulted in a slow reaction, even though many people were convinced that the mills were causing the problem. The mills were finally shut down, and the weather began to improve on October 31, when a southwesterly wind in advance of an approaching cold front began to disperse the deadly smog. Light rain associated with the cold front further cleared the air on November 1, but the tragic results of the fog and smog would be remembered forever.

Did the Weather Kill a President?

Although we like to think of history as being composed of cold, hard facts, my eighth-grade history teacher used to hint about how pliable history might be. Perspective, he taught, is as important to understanding history as is knowledge, and from a weather perspective, the widely held belief that the death of President William Henry Harrison, the ninth president of the United States, was caused by the weather is most likely a historical contrivance.

William Henry Harrison was inaugurated on a cold and wet early March day in 1841, but he donned no overcoat, hat, or gloves as he gave an incredibly long inaugural address. The nearly 8,500-word speech took close to 2 hours to deliver and still stands as the longest inaugural address on record. Exactly 1 month later, on April 4, Harrison died of pneumonia, and the death of the president with the shortest time in office was blamed on the inaugural-day weather.

While the possibility of a lowered resistance to disease created by his being cold exists (see "News Flash: Cold Air Doesn't Make You Sick" in Chapter 8), some historic records indicate that he showed no signs of sickness until a full 3 weeks later. Given that time frame, it's unlikely that the illness was directly related to the weather.

It's a better story, though, if we make the claim that the president, who spoke longer than any president before or after him on a cold, damp day, brought about his own demise.

The Flood City

Johnstown, Pennsylvania, a small valley city in western Pennsylvania, was important in the 1800s and early 1900s in terms of water travel, railroad travel, and iron production. Sadly, though, the city will always be defined by its tragedies, not its accomplishments. Johnstown is the site of multiple historic floods.

Floods Before and After 1889

Other noteworthy floods in Johnstown include an 1833 flood, when water levels rose to nearly 27 feet above flood stage, and a devastating St. Patrick's Day flood in 1936, which that resulted in $42 million in property damage. Fortunately, a better warning system had been established by the time, so fewer residents perished (25). A 1977 flash flood, which resulted from nearly stationary thunderstorms producing as much as a foot of rain in less than 24 hours, left much of the city under 6 feet of water.

The most deadly of these floods, which occurred on May 31, 1889, still stands as one of the worst natural disasters in the nation's history. A strong, slow-moving storm from the west dumped 6 to 10 inches of rain in just 24 hours and poured torrents of water into the waterways; this, combined with questionable engineering and a dam of suspect quality, resulted in disaster. The South Fork Dam, which sat 14 miles upstream of the city, burst. The ensuing wall of raging water reached heights of 60 feet as it raced through the narrow valley, destroying everything and, tragically, everyone in its path, claiming 2,209 lives.

Freedom from Heat and Humidity

Summer in Philadelphia is notorious for heat and humidity, and the spring and early summer of 1776 was hotter than average;

however, the weather on the Fourth of July was delightful, with lower-than-average temperatures and low humidity, at least according to the person who was most like a meteorologist of his time: a man named Thomas Jefferson.

The days immediately before the signing of the Declaration of Independence were still hot. July 1 was hot and steamy with a heavy thunderstorm, but the thunderstorm did little to relieve the heat—and most likely increased the humidity (see "It's Always Less Humid After a Thunderstorm" in Chapter 2). A thunderstorm on July 2, however, was obviously accompanied by a cold front, since the temperatures on July 3 were about 10°F lower, and by July 4, it was delightful.

Jefferson, who kept detailed weather records (as if he didn't have enough to do), indicated a 6:00 a.m. temperature of 68°F, a 1:00 p.m. temperature of 76°F, and a 9:00 p.m. temperature of 73°F. Based on our modern averages, the approximate high of 79°F or 80°F (with low humidity) was about 6°F below normal. Without the benefit of air-conditioning (Franklin should have invented that, too!), it was much more comfortable than previous days and the typical summer heat.

It was one of those spectacularly bright, blue days that makes it seem as if anything were possible.

Galveston Hurricane

Hurricane Katrina (see "Katrina" later in this chapter) is the defining natural disaster of the early 21st century for the United States, but it was another Gulf of Mexico hurricane, an unnamed storm (unnamed because it occurred before tropical storms and hurricanes were assigned names), that brought unimaginable tragedy to Galveston, Texas, at the start of the 20th century. On September 9, 1900, a major hurricane resulted in the death of 6,000 to 8,000 people as it plowed through southeastern Texas.

Meteorology was in its infancy in 1900; without satellite photographs or computer models to warn of the storm, residents had little idea of the peril awaiting the city—even though meteorologists, who monitored the sky, wind, and waves, believed a storm was approaching. Weather Bureau meteorologists dutifully recorded observations before and during the storm—at least until the weather instruments were blown away by a sustained wind estimated to have peaked at 130 mph.

As is often the case with hurricanes, the storm surge was the most damaging part of the storm. Galveston, which is located on an island in the Gulf of Mexico, was flooded by one that reached at least 20 feet. Thousands of people who'd fled beach communities for stronger buildings farther inland were saved; however, thousands of those who'd stayed died. The storm was the impetus for building a 17-foot sea wall to protect the city from future storms.

Inaugural Melodrama

Weather has always been used as symbolism in Hollywood, so it seemed appropriate when the weather showed a tendency toward drama for actor-turned-politician Ronald Reagan.

His first inauguration, on January 20, 1981, was the warmest

Hurricane Ike

Hurricane Ike, which blasted through the Texas coast on September 13, 2008, with estimated sustained wind speeds of 125 mph, was a devastating storm. Ike resulted in dozens, if not hundreds, of deaths (the official death toll has not been determined) and roughly $27 billion worth of damage in the United States alone. The storm surge, however, was lower than the original forecast of 25 feet. It peaked at about 13.5 feet on Galveston Island, with wind waves on top of the surge crashing over the sea wall, causing some flooding; however, the smaller-than-forecast surge prevented the city from having even greater devastation.

inauguration to date (at least since official records have been kept), with a noon temperature of 55°F. This was in sharp contrast to the 7°F reading (windchill temperature of 10°F to 20°F below zero) that greeted President Reagan on January 21, 1985, which was the coldest inauguration in history. I guess that temperature didn't technically *greet* Reagan, since the inauguration was held inside due to the extreme cold.

From warmest in history to coldest in history—seems like overacting to me.

Incredible Fall Snowstorm

Extreme but isolated snowstorms occur downwind of the Great Lakes every winter as cold air passes over the warm lakes, resulting in tremendous accumulations of snow; however, it's hard to imagine any such lake-effect snowstorm being more impressive than the one in Buffalo, New York, in the middle of October 2006. That's right, one of the worst snowstorms in Buffalo history occurred on October 12–13, which is when fall foliage is typically near its peak color.

When a very cold early-season air mass approached western New York, the temperature of Lake Erie, after a mild fall, was a balmy 62°F. Even though cold air passing over a warm lake is the recipe for massive lake-effect snow, forecasters generally believed the cold air would not be cold enough or deep enough (meaning a thick enough layer of cold air) to withstand the warming associated with passing over the relatively steamy water. Some wet snow was expected, but forecasters believed that much of the precipitation would fall in the form of rain—lake-effect rain. Heavy lake-effect snow just could not occur this early in the season, especially since snow accumulations of greater than 2 inches had only occurred twice in October.

Well, not only was the air mass cold enough to produce snow, but it also took full advantage of the contrast between the cold air and the warm lake to produce a tremendous amount of it, much of it accompanied by snow-muffled thunder. The nearly 2 feet of snow (22.6 inches) that accumulated at Buffalo's airport was the seventh-largest snowstorm in its storied history of huge snowstorms, and much of western New York received between 1 and 2 feet of snow.

The snow was heavier and wetter than the typical lake-effect snow because the lake waters were so warm (it was nearly a rainstorm), and many of the trees were in complete foliage. An

Strongest Atlantic Basin Hurricanes on Record (by Pressure in Inches of Mercury)

Katrina (2005)	26.64
Allen (1980)	26.55
Rita (2005)	26.43
Unnamed (1935)	26.34
Gilbert (1988)	26.22
Wilma (2005)	26.05

incredible amount of damage was done to trees and power lines because the leaves caught the heavy, wet snow, bringing limbs down. Nearly a million western New Yorkers were without electricity, some for about a week.

Katrina

No discussion of historic United States weather events would be complete without a discussion of Hurricane Katrina, one of the nation's worst natural disasters—although, to be fair, it was partly a man-made disaster because the most tragic part of the storm was the result of a failed man-made levee system. Words here cannot describe the type of suffering the storm caused—and is still causing—to those who lost loved ones and homes or were forced to move away from a city they loved. I'll focus on the meteorology.

Since any Hurricane Katrina discussion understandably centers on New Orleans, it has been forgotten by many of us that Katrina initially made landfall in Florida, moving across south Florida on August 25, 2005, just 2 days after a tropical depression had formed in the Bahamas. The hurricane caused serious flooding and more than a dozen fatalities.

Katrina had weakened to a tropical storm by the time it exited from the west coast of Florida, but after the storm moved into the open waters of the Gulf of Mexico, Katrina quickly re-intensified, becoming a category 5 hurricane (the strongest category) and the sixth most intense hurricane in Atlantic Basin history (admit it: you thought it was the strongest). The storm weakened to a category 3 by the time it made landfall in southern Mississippi (admit it: you thought it made landfall in Louisiana) on August 29.

Minor Storm; Major Rain

As a hurricane, Agnes was probably as forgettable as they come, making landfall in the Florida panhandle as a weak category 1

2005 Hurricane Season

The 2005 hurricane season produced 27 named storms (a record), along with seven major hurricanes (category 3 or higher). Four of these major hurricanes—Dennis, Katrina, Rita, and Wilma—made landfall in the United States, including Wilma, which was the most intense Atlantic Basin hurricane and devastated much of south Florida in late October. The storm killed over three dozen people and caused approximately $20 billion in damage.

storm in the middle of June 1972. After the storm weakened to a tropical depression while moving through the Southeast, Agnes regained tropical storm strength off the Carolina coast and moved northward, bringing heavy rain to the coastal Middle Atlantic region. Agnes was then pulled westward into Pennsylvania, where it merged with a slow-moving, upper-level storm system. When the remnants of a tropical system merge with a non-tropical system, rainfall can be historic, which was the case in Pennsylvania during a period that lasted nearly a week.

A large portion of the state, mainly over the Susquehanna River Valley, had over 10 inches of rain, with some locations approaching 20 inches. The resulting flooding remains the worst in history in many locations and has been called a 500-year event, meaning it's the type of weather event that should happen only once every 500 years. Four dozen deaths were blamed on Agnes, and billions of dollars worth of property and crops were destroyed.

Agnes is also proof that any tropical storm system is a potentially historic weather event, one destined to be talked about for generations to come.

Southern White Christmas

While the idea of a white Christmas is greatly romanticized in this country, residents in the Deep South had rarely seen snow on Christmas, other than on television, in movies, and on Christmas cards—until the historic snow on Christmas Eve 2004, that is.

An arctic cold front moved through south Texas and south-

Devastating Tropical Storm

We will always remember powerful hurricanes such as Katrina and Andrew, which caused devastation of epic proportions to U.S. cities; however, some of the truly historic tropical events were never hurricanes.

Tropical Storm Allison was one such storm, being the most costly of all tropical storms to affect the United States. Allison never had sustained winds of greater than 60 mph when it was a tropical storm over the western Gulf of Mexico; in fact, Allison was a tropical storm for a scant 12 hours before making landfall in southeastern Texas on June 5, 2001. The storm had no storm surge and produced tides of only a couple of feet above normal along the Texas coast; however, this relatively innocent-sounding tropical storm remained over southeastern Texas from June 5 through June 9, producing the worst flooding in Houston history.

Over that 5-day period, bands of torrential rain, falling at a rate of 4 inches per hour for a time, resulted in 26 inches of rain in 10 hours—that's almost as much rain as Detroit typically receives in an entire year, all in less than half a day. Numerous locations in and around Houston received over 2 feet of rain during the deluge. Nearly two dozen people were killed by the flooding, and property damage was estimated at $5 billion.

ern Louisiana late in the day on December 22, setting the stage for a bitter blast of winter cold. High temperatures dropped from the 70s and lower 80s on December 22 to the 30s and lower 40s on December 24. A storm system then moved from the Desert Southwest to just south of the region, throwing moisture from the Gulf of Mexico into the cold air and producing a remarkable snowfall.

Historic Southern Snowstorms (in Inches)

El Paso, Texas	22.4	December 13–14, 1987
Las Vegas, Nevada	9.0	January 4–5, 1974
Birmingham, Alabama	17.0	March 12–13, 1993
Tallahassee, Florida	1.0	December 22–23, 1989
Phoenix, Arizona	1.0	January 20, 1933

Some southern cities had just a touch of Christmas magic—Houston had 1 inch, marking its first white Christmas on record; New Orleans had 0.7 inches; and Brownsville had 1.7 inches—but it was an all-time record-breaking snowstorm for many others. Victoria, Texas, received a whopping 13 inches of snow, and Alice, Texas, received a foot of snow. Even typically tropical Corpus Christi had just over 5 inches.

Storm of the Century

I'm not one for hyperbole, but the so-called Storm of the Century (or Superstorm) in March 1993 (March 12–15) was easily the most memorable storm in my more than 20 years of forecasting. While it's often referred to as the Blizzard of '93, the Storm of the Century is a better description because the storm was more than just a blizzard.

The blizzard—and this was a true blizzard (see "Every Heavy Snowstorm Is a Blizzard" in Chapter 2)—part of the storm was incredible, with tremendous snowfall from the Deep South (6-foot drifts in Alabama) to New England. Many locations received over 2 feet of snow, including areas from Connecticut to Tennessee. Snow and rain, accompanied by powerful wind, pounded the major cities along the coast.

The merger of three separate upper-level storm systems fueled the storm, and the storm system was well forecast by the computer models several days in advance, which is unusual today but was exceptional in 1993. Not only did the storm produce massive snow, but it also produced tremendous wind, including a 70 mph wind gust in New York City.

One devastating part of the storm—sometimes omitted from the discussion because the blizzard aspect was so impressive— was the outbreak of damaging thunderstorms and tornadoes that ravaged Florida. Numerous locations reported wind gusts in excess of 100 mph in thunderstorms, and 11 tornadoes touched down. The death toll was in the dozens, making it one of the worst tornado outbreaks in the state's history.

In total, fatalities were estimated to be at 300—the inexact number the result of the different ways that storm-related deaths are calculated. (Some estimates include those who had heart attacks while shoveling snow, for example, while other estimates don't.) Over a million people were left without electricity, thousands of cars were stranded on highways, and airports from the Deep South to New England were closed.

Super Tornado Outbreak

Spring is notorious for violent weather because the last storms of an active winter season collide with the growing heat and humidity of the approaching summer, and the so-called Super

Tornado Outbreak of April 3–4, 1974, was one of the most graphic displays in American history.

A strong, late-season storm hit northern California on April 1, and when the storm moved into the Midwest a couple of days later, it collided with the warm, humid air that had originated in the Gulf of Mexico. The result was over 148 tornadoes in 13 states (from Illinois eastward to North Carolina, and from southern Ontario southward to Alabama) during a 16-hour period spanning 2 days, resulting in 330 deaths, 5,484 injuries, and property damage of approximately $600 million. In total, it's estimated that nearly 28,000 families were affected by the storm.

The tornado outbreak occurred on the heels of a strong La Niña, and a La Niña winter often means an active tornado sea-

Great Blizzard of '88—1888, That Is

While the 1993 storm was the most impressive that I'd ever seen, that might be because I wasn't around in 1888, when an incredible blizzard crippled the northeastern part of the country. This storm, also in March (11–12), produced higher snow amounts than the blizzard of 1993 in some areas, including 50 inches in Connecticut, and the combination of snow and wind resulted in unimaginable drifts of 25 to 40 feet.

Even if those numbers have been exaggerated over the generations (or not measured with the precision of modern observations), the combination of wind and snow was clearly more intense in the Northeast than it was in the 1993 storm. The 1888 storm did not, however, bring snow to as large an area or produce the number of tornadoes as did its modern counterpart.

son in the spring. La Niña seasons, which magnify what is already a typically sharp contrast between the seasons, are characterized by a more-active-than-normal spring storm track, such as the strong, late-season storm of this particular year.

Walking on Water

Winters in Valley Forge, Pennsylvania, were very cold in the 1770s; otherwise, George Washington's soldiers would not have roasted and eaten their boots—they'd have ordered in cheesesteaks instead. Since they didn't, the only logical conclusion is that it was so cold that no one would deliver.

Complete weather records for the 1770s don't exist even for a historic city such as Philadelphia. (My theory is that Benjamin Franklin was too busy attending parties in France to keep weather records, which was supposed to be his charge in Philadelphia.) Our historic records, however, do indicate that the weather did, indeed, play a significant role in the Battle of Trenton, a key battle in the Revolutionary War.

A large, flowing river, such as the Delaware, does not usually freeze at all during the heart of winter, which is January and February, including winters with periods of intense cold. When such a river has ice jams in December, as it did in 1776, not only is it an intensely cold start to winter, but it's intensely cold with little or no breaks. For a meagerly supplied army in 1776, the hardship had to have been more than we can imagine—and on top of that, the plan was to cross the river and attack the enemy on Christmas night (December 25).

A raging storm met these remarkable, ill-equipped heroes as they marched forth and then crossed a swollen river filled with sheets of ice, with worn-out leaky boats transporting frightened horses. A wind-driven rain, which changed to a pelting sleet and snow as they traveled eastward, drenched their tattered clothes, and although I'm sure that it didn't seem like it at the time, this Nor'easter was more of a blessing than a curse. Not only did the weather mean that the surprise tactic was indeed a surprise, but the sound of the wind and pounding precipitation dampened the sound of Washington's army.

Washington didn't literally walk across an ice-covered river, but in retrospect, it seems as if he and his soldiers could have walked on water.

The Year Without a Summer

Complaints about the weather are as popular as popcorn at a double feature, but compared to the summer of 1816, we have nothing to complain about.

Two snowstorms occurred in eastern Canada and New England in June, with a killer frost in the Middle Atlantic region. Another killer frost occurred in early July, when ice and snow was, incredibly, reported on rivers in Pennsylvania. Another round of cold air brought a third killer frost to much of New England during the latter part of August, effectively ending the growing season, one that had been a disaster to begin with. The results of the "year without a summer" were tragic: It's estimated that 1,800 people died of exposure, with widespread famine and hunger.

Without the worldwide communication and observation we have today, it's difficult to know with certainty the cause of the extreme weather; however, a massive volcanic explosion (Mount Tambora, in Indonesia) is generally believed to have been the cause. Large volcanic eruptions shoot incredible amounts of ash and soot tens of thousands of feet into the air, blocking the sun that warms the Earth and drives the normal weather processes. Some scientists believe that threats such as large volcanic eruptions pose a greater threat to the world than does global warming.

When looking at an event such as this from 21st-century America, it's important to keep it in the perspective of the time. Fruit, vegetables, and meat were generally grown locally, so residents in New England and the Middle Atlantic regions depended on local farmers. The East was, of course, the most populous area (places such Indiana, Illinois, Michigan, Missouri, Texas, Florida, and California were not yet states), so assistance was largely unavailable.

5

Field of Dreams

SPORTS-RELATED WEATHER

AS SOMEONE WHO HAS ALWAYS been interested in the weather *and* sports, my greatest sports memories often involve the weather, typically how it affected a sporting event in a unique way. It is not surprising, then, that I'm not a fan of domes, stadiums with re-tractable roofs, heated fields to melt snow, and turf that can drain 10 inches of rain in an hour, since those take some of the fun out of the games.

I vividly recall a meaningless regular-season Major League Baseball game in the 1970s when an outfielder for my then-beloved Pittsburgh Pirates was unable to catch a fly ball in Shea Stadium in New York because of the fog. The player stared help-lessly into the misty sky as the baseball landed with a plunk only a couple of feet from him. What I don't recall is the thousands or tens of thousands of times (hey, I watched many games when I was young) when the baseball landed safely in an outfielder's glove because the weather was not a factor.

For most of us, it's more memorable when the weather has an effect on a landmark game, such as a playoff game in a professional team sport or the Winter or Summer Olympics. This chapter includes some of those events, but it also includes a few for which the weather, more than the game, was the story.

This section has a football bias—not because I enjoy football more than other sports but because it's the one professional team sport regularly played in all types of weather, and the most important games are played during the heart of winter, which is when the weather is the most interesting.

Fog Bowl

The weather in Chicago on December 31, 1988, was typically cold, with morning temperatures in the teens, and when a weak easterly breeze brought milder, more moist air from Lake Michigan over the cold, snow-covered ground, a dense fog quickly formed. None of that is particularly noteworthy except that an NFL playoff game between the Philadelphia Eagles and the Chicago Bears was being played at Chicago Stadium when the fog arrived—well, at least attempting to be played.

This wasn't your average fog, the type that makes it a little more difficult to see what's half a mile ahead when driving down the highway. The visibility was reduced to under 10 yards at times, which is a serious problem with a 100-yard field. I remember watching the game on television—and I'm using the word *watching* loosely. Players couldn't see the field markers. Passes would bounce off receivers who didn't see the ball coming. Punts would rise into the mist only to drop 10 or 15 yards away from would-be punt returners. Fans in the stands couldn't see what was going on. The announcers in the booth were guessing what was happening on the field.

Chicago managed to win the game by a score of 20–12, but this is a good example of how normally reasonable people often become unreasonable when talking about sports. A friend and Philadelphia sports fan insisted that the game should have been stopped because of the fog, even though football does not generally stop on account of weather (except for rescheduling games because of natural disasters, such as hurricanes, or pulling the players from the field during a thunderstorm). Philadelphia, the team trailing in the game during the worst part of the fog, was at a disadvantage because of the fog, my friend argued.

Unless the fog magically appeared when Philadelphia was on offense and disappeared when Chicago was on offense, I'm not sure how that argument holds water—or should I say *fog*.

Hockey—in a Fog

Fog can form in a variety of ways, one of which is when warm, moist air crosses snow- or ice-covered ground, but it can also form inside of an aging arena where hockey is attempting to be played.

The original Boston Garden, not the modern building with the same name, was built in 1928, and one of its many design flaws was a lack of air-conditioning. The warmth and humidity associated with thousands of people jammed inside of a building during a relatively warm evening often contrasted enough with the cold air from the ice to produce fog over the rink. Because the hockey playoff season extends through April and May into June, when the buildings are warmer on the inside, the fog problem was at its worst when the games meant the most.

A particularly noteworthy fog-filled game was part of the Stanley Cup Finals in late May 1988. The fog was so thick that the players—reminiscent of the days when players, not the Zamboni, resurfaced the ice between periods—were instructed to skate in circles around the nets in order to break up the fog, but in the second period, the electricity went out (unrelated to the fog problem), and because the generators could only last for about 30 minutes, the game was postponed.

The Boston Garden (may it rest in peace) certainly wasn't the only ice arena to have fog problems, but it was the most famous.

Ice Bowl

Football, cold weather, and the Packers—it's a classic combination for sports fans. In fact, the expression "frozen tundra of Lambeau Field" being said in an overly deep voice in an attempt to imitate John Facenda, the baritone-voiced original narrator of NFL films, is a football cliché.

Green Bay seems to have the reputation as the coldest of all NFL cities, but meteorologically speaking, it's not significantly colder than other NFL cities—ones never confused with Miami in December—including Minneapolis (at least before they moved into the giant plastic bag—see "Much Adome About Nothing?" later in this chapter), Detroit, Chicago, Buffalo, Foxboro (home of the New England Patriots), Cleveland, Cincinnati, and Pittsburgh. The combination of important games played in the bitter cold by the historic franchise has sealed its reputation, though, and the most famous of all cold-weather games is affectionately called the Ice Bowl, played on December 31, 1967.

With players and coaches such as Bart Starr, Vince Lombardi,

Favre's Last Lambeau Hurrah

Since his career started in 1992, no one has represented the Green Bay Packers and Lambeau Field better than Brett Favre, so it seems only appropriate that his last playoff run as a Green Bay Packer was filled with harsh-weather football games.

When the Packers hosted the Seattle Seahawks on January 12, 2008, snow fell heavily at times, especially in the fourth quarter, when Green Bay had a solid lead in the game. Even though snow accumulation and temperatures were not extreme (less than 1 inch of snow officially accumulated in Green Bay, and the temperature was just a couple of degrees below freezing), the field was covered in snow, the ball was slippery, and if you looked carefully through the snowflakes, you could almost see the ghost of Vince Lombardi standing on the sidelines.

Eight days later, Favre and the Packers lost a heartbreaking NFC Championship game to finish the season, but by then, arctic air had arrived in Green Bay. The game-time temperature was −1°F on January 20. The New York Giants won the hard-fought game when a rock-hard football was kicked through the uprights for a game-winning field goal.

Tom Landry, and Ray Nitschke involved in a league champion-ship game with a kickoff temperature of a football-numbing −13°F and a windchill of −35°F, it's not surprising that the game is one destined to be remembered for generations. The Packers won by a score of 21–17, with a late touchdown scored by Starr.

The unbelievable cold not only affected the players but also affected the referees, who couldn't use their frozen whistles, and the marching band, which couldn't perform.

It's Not the Steroids; It's the Humidity

I sometimes feel bad for humidity. It gets blamed for everything from uncomfortable heat (see "It's Not the Heat; It's the Humid-ity" in Chapter 2) to cookies and bread not rising (I'll refer you to my mother on that one) to general aches and pains. In most instances, the blame is warranted; however, when humidity is blamed for fewer home runs in baseball because "the air is heavy," then I have to jump in and defend the humidity.

Not only does higher humidity *not* result in fewer home runs, but it actually allows a baseball to travel farther, increasing the likelihood of home runs. The molecular scoreboard gives the details: The molecular composition of a unit of air is lighter when it's humid than when it's dry (see "The Air Is So Heavy [When It's Humid]" in Chapter 2). The air feels heavier because it makes us feel uncomfortable when the sweat from eating our $8 nachos doesn't evaporate into the air, but the air is, in fact, lighter than it is on a day with low humidity.

Hot air is also lighter than cool air, so a baseball travels most effectively on a hot, humid afternoon. Perhaps it's not the steroids that have caused the surge in home runs in recent decades but global warming.

Humidors Are Not Just for Cigars

While hot, humid air is certainly more conducive to home runs than cool, dry air, another significant factor influencing how far a baseball will travel is a stadium's altitude. In the altitude competition, Denver's Coors Field wins every year because it's the highest Major League Baseball stadium. In fact, a row of purple seats inside the stadium marks an elevation of 5,280 feet (or 1 mile) above sea level.

While we've known that light air (fewer air molecules) at higher elevations allows a baseball to fly farther than in lower elevations, a study by the University of Northern Colorado quantified this knowledge. The study shows that the same hit ball would travel 400 feet in Miami and 420 feet in Coors Field, and if you think about how many balls are caught within 20 feet of an outfield wall, then you have some idea of what a difference that would make.

In order to attempt to keep Coors Field in the same ballpark as far as home-run production is concerned, Major League Baseball gave the Colorado Rockies permission to store baseballs in a humidor, which adds moisture to the balls. While humid air is lighter than dry air, a humid ball is heavier than a dry ball—because of the weight of the added moisture. This has balanced the home run playing field to some degree.

NOTE: The light air effect applies to golf balls as well as baseballs, so golfers having a difficult time getting distance from drivers might want to consider a summer golfing vacation in the Rockies!

It's Raining Basketballs

Professional basketball is played indoors, so weather-related cancellations and delays are as rare as a leaky roof in an $800,000,000 arena, but it has happened.

Southern California rainstorms were rare in 2008 (it was a dry winter even by Southern California standards), but a storm on January 27–28 produced about 1.5 inches of rain in Los Angeles, along with a leaking roof on the Staples Center. The result was a 12-minute rain delay of the game between the L.A. Lakers and the Cleveland Cavaliers.

It's Snow Problem

The 1980 Winter Olympics in Lake Placid, New York, is probably best known for the Miracle on Ice, which is the gold-winning performance by the U.S. Men's Hockey team. (For the record,

Snowflakes as Large as Basketballs

A National Basketball Association game was canceled because of a blizzard on December 20, 1996, but this time, the weather stayed where it belonged—outside. Nearly 2 feet of snow in Denver resulted in the cancellation of a game between the Phoenix Suns and the Denver Nuggets. Both teams were in the city, but road conditions were too hazardous for fans. The Suns, by the way, were so delayed by the weather that they nearly missed a game the following night in Phoenix.

the U.S. team beat Finland, not Russia, in the gold medal round.) For fans of weather, though, the 1980 games were also known for something else—the first use of artificial snow in Olympic competition.

Nearly every ski resort in the United States has had its share of poor winters for skiing, even those located in the cold climes of the Adirondack Mountains of upstate New York. That's where Lake Placid, New York, is located, and the 1979–1980 winter was one of those years—exceptionally mild and generally snowless. (It was not an El Niño winter, for those of you who believe that such weather has to be associated with El Niño.)

A lack of snow is always bad news for ski resorts, but it's a potential source of national embarrassment when the best winter athletes in the world are gathering for the Olympic Games. The largest production of man-made snow in history allowed ski competitions to go on, marking the first use of artificial snow in Olympic history (even though the snowmaking had been in use for many years).

Jumping from the Ice Box to the Freezer Bowl

While the Ice Bowl (see "Ice Bowl" earlier in this chapter) is the most famous cold-weather game in NFL history, another playoff game was played in remarkably cold weather on the frozen turf of Riverfront Stadium, in Cincinnati, Ohio. The game, sometimes referred to as the Freezer Bowl, was played on January 10, 1982.

An arctic high pressure system brought the bitterly cold air; the game-time temperature was −9°F. Ken Anderson, the

Original Lake Placid Snow Problems

> The first time that a lack of snow threatened the winter games in Lake Placid was the 1932 Lake Placid games, which were also nearly snowless and at a time when making snow was not an option. (There is no truth to the rumor that the country, in the Depression, was too poor to afford snow.) A couple of snowstorms just before the Olympics saved the games.

Cincinnati Bengals quarterback, must have had plenty of coffee that day (he was famous for coffee commercials in the 1970s), leading the hometown Bengals to a 41–10 win over the San Diego Chargers, who, I'm sure, were longing for their allegedly perfect climate (see "San Diego Has the Perfect Climate; It's Always Sunny and 70°F" in Chapter 2).

A punter usually has the easiest job on the field (unless one of those really big, fast guys happens to run into him while his foot is above his head), but this was one of the worst days in NFL history to punt. Not only does extreme cold keep a ball from flying as far (cold air is heavier than warm air), but the ball had to have been nearly rock hard from the cold. In addition, the 27 mph wind brought windchill temperatures down to an unmerciful −40°F (which is also −40°C, by the way), which interfered with the direction and distance of the punts.

Much Adome About Nothing?

With the tendency toward domed stadiums in recent decades, forecasting the weather for major sporting events is often as simple as reading a thermostat; however, the Metrodome in Minneapolis, Minnesota, is one of the few domed sporting arenas in which the weather is rumored to have an effect on the games being played inside.

The Minnesota Twins franchise (Major League Baseball) was fairly successful during the second half of the 1980s and the first part of the 1990s amid frequent rumors that the maintenance staff of the Metrodome (the inside of which looks like a giant plastic garbage bag) would alter the air flow of the stadium by adjusting the air conditioner. In short, more air-conditioning when the home team was at bat meant a stronger wind blowing out, which could result in more frequent home runs.

Rumors were so rampant that studies were conducted by a Minnesota university to see whether the indoor wind could be manipulated by the air-conditioning system. The reports were inconclusive, but according to an interview with the *Minneapolis Star-Tribune*, the then-superintendent admitted to trying to influence the game.

It gives a new definition to *home-field advantage*.

October Classic Snowflakes

Fewer baseball games have notable weather stories than football games because baseball isn't generally played during inclement weather—baseball is considered too much a finesse sport. A baseball game might continue through a light rain or drizzle, but games played in major snowstorms, bitterly cold outbreaks, and horrendous rainstorms are generally limited to football. With

There's No Place Like Dome

Regardless of whether a man-made wind influenced play, the weather has affected the Metrodome in other ways over the years—and with an air-supported fabric roof, it's not surprising. The roof collapsed (deflated) from the weight of a heavy November snowfall, and a 1986 thunderstorm ripped a hole in the roof while a baseball game was going on.

the expansion of the length of the season, though, surprisingly bad weather occasionally spills into the so-called October classic. The 1979 season is a good example.

Inclement weather was not out of the ordinary in the seven-game series between Pittsburgh and Baltimore, with one postponement and one game played in the rain, but an unusually cold storm system brought the earliest recorded snowfall in Baltimore history on October 10, 1979, which was the date of game one of the series. The official snowfall was just 0.3 inches, but it was in sharp contrast to a typical October 10 in Baltimore, which has an average high temperature of 69°F.

The snow melted by scheduled game time, so the game was played, but a sloppy field meant sloppy play. The teams combined for six fielding errors.

Outdoor Hockey

Professional hockey, of course, is played in weather-protected indoor arenas all across the globe, so weather-related game cancellations are generally related only to the safety of fans and teams; however, a recent development in the National Hockey League (NHL) has started to change that. The league has started playing the occasional game outside, replicating the millions of games that have been played on frozen ponds, lakes, and make-shift outdoor rinks in cold climates for generations—the field of dreams for many Canadian youngsters.

The first such NHL outdoor game in the United States was in Buffalo, New York, on January 1, 2008, between the Pitts-burgh Penguins and the Buffalo Sabres. While the Penguins won the game on a shoot-out goal by superstar Sidney Crosby, the weather was the star of the show. The combination of wind and temperatures in the teens resulted in windchill temperatures below zero, and frequent snow flurries filled the sky. None of this affected attendance, which was an NHL-record 70,000.

The possibility of cancellation (due to major snowstorms, rain, or fog), the limited number of cities that have the weather to make outdoor games feasible (it's tougher to maintain an outdoor rink in Phoenix or Tampa Bay), and the expense of game production will most likely prevent an abundance of such games in the future.

Philadelphia Curse Washed Away

It took 25 years and a weather-delayed baseball game spanning 3 days, but a chilly October rain finally washed away Philadel-phia's professional sports curse. With the Philadelphia Phillies leading the Tampa Bay Rays three games to one in the best-of-seven World Series in 2008, the potential championship-deciding game was played on a stormy Philadelphia night.

Like Kissing an Ice Sculpture
of Your Sister

Professional sports tends to be the precedent setter; however, the inspiration for NHL outdoor games may have been an outdoor college hockey game between rivals Michigan State and Michigan.

On October 6, 2001, temperatures were in the 30s in Lansing on the day of the so-called Cold War, which is chilly for October but certainly comfortable enough for players on the ice. The main factor was the wind, which increased the threat of dehydration. The game was played in front of a record-breaking crowd of over 75,000 screaming fans.

The game ended in a thrilling tie (3–3), and a tie in hockey is often compared to kissing your sister. In this case, it's more like kissing an ice sculpture of your sister.

Rain didn't begin until the fourth inning, when Philadelphia had a two-run lead. While only a couple of tenths of an inch of rain fell during the game, the rain created puddles in the infield, making the base paths muddy, and a whipping wind made catching the ball difficult and added to the chilly conditions. Temperatures dropped into the 40s, which would have been good weather for the Eagles versus the Buccaneers (the two professional football teams from those cities), but it wasn't good weather for baseball.

When the wind and rain interfered with a Philadelphia player's catching an easy pop-up, resulting in Tampa scoring a game-tying run, it appeared as if Philadelphia's curse would continue, perhaps for another quarter of a century or so. Play was suspended in the sixth inning with the game tied, though, which was the only time that an ongoing World Series game was suspended by weather.

When the game resumed on a brisk evening two nights later (the slow-moving storm resulted in no baseball being played the next day), Philadelphia managed to regain the lead and win the game—the curse of Philadelphia sports teams was finally over.

Rain, Rain—I'll Make You Go Away

"Everyone talks about the weather, but no one does anything about it" is a famous weather quote attributed to Mark Twain.

Philadelphia Weather Problems Decades Earlier

The 2008 World Series was not the only important Philadelphia baseball game affected by the weather. The final game in the National League Championship game in 1977, played in Philadelphia on October 8 between Philadelphia and Los Angeles, was also played in the rain following a 2-hour rain delay before the first pitch.

Los Angeles won the game to advance to the World Series. While Philadelphia fans might like to blame the weather for their loss, squandering a two-run lead in the last inning of the previous game had more to do with the loss than did the rain.

Others, though, give Charles Dudley Warner credit for the statement. Rather than debating who said it first, China claims to have the power to heed the advice, at least based on its 2008 commitment to stopping the rain from falling from the sky if it threatened during the opening or closing ceremony of the Summer Olympics.

An approaching thunderstorm did, in fact, diminish before reaching the stadium during the opening night ceremonies, prompting officials to claim that their massive cloud-seeding effort was a success. I wasn't in China (I don't even like to drive to the grocery store after dark), but it's a matter of debate as to whether the government's cloud-seeding program had any effect.

Roast a Chicken on the Football Field

We've already talked about the myth of being able to cook an egg on a sidewalk (see "Hot Enough to Cook an Egg on the Sidewalk" in Chapter 2), but when talking about artificial turf, skip the egg and cook the whole darn chicken.

Professional sports stadiums built from the late 1960s through the 1980s (roughly speaking) were often equipped with the then-trendy artificial playing surface (often referred to by the brand name AstroTurf). The advantages included being able to have a green turf in December in places such as Buffalo, and easier drainage and maintenance. As it turned out, more athletes were injured on the artificial grass.

Planting Rain?

Cloud seeding is one of the most common forms of attempted weather modification, and while the theory is sound, the results have been inconclusive.

Water remains in a liquid state at temperatures well below freezing when condensation nuclei (microscopic particles onto which the water can freeze) are lacking (see "Precipitation Formation" in Chapter 1). Cloud seeding is dropping artificial condensation nuclei, typically in the form of silver iodide, into a cloud that contains supercooled water in the hopes of kick-starting the precipitation process. This is referred to as *static cloud seeding.*

Cloud seeding can also be done to a cloud that is already producing precipitation. The theory is that the latent heat, produced by the additional freezing caused by adding condensation nuclei, will increase the heat-driven updraft producing the cloud. This is supposed to make the cloud rain itself out more quickly—and save important events, like the Olympics, from being rained out.

Scientific results related to cloud seeding have shown spotty success at best, and it's difficult to believe that cloud seeding can compete with the tremendous forces of nature.

It's much hotter on pavement than it is on grass (perhaps 40°F higher), but the difference in heat between grass and artificial turf is unfathomable—and potentially dangerous to athletes. According to a Brigham Young University study, the temperature on synthetic turf (can you say oxymoron?) was found to be 86.7°F higher than on grass at an elevation of 5 feet 6 inches above the ground, which is generally—within approximately a foot—the height of athletes, who need to breathe and perform in this anomalously hot air. In the study, irrigation (spraying water) of the turf significantly cooled the air at the same height, but the cooling was temporary; within 20 minutes, the temperature had soared back to its original 164°F.

Recent decades have seen many new stadiums being built, largely out of the need to generate more revenue; however, the majority of these stadiums have gone back to good ol' grass. The only disadvantage is that the chickens will have to be cooked on the rotisserie instead of at second base or the 50-yard line.

Smog Time

The 1984 Olympics was probably best known for the Soviet-led boycott; however, there may be some confusion about the reason for the boycott. Political dispute was the officially reported cause, but it was probably because the whiny Olympic athletes were afraid of the smog. You know how those prima donna athlete types are: "I'm having trouble breathing." Whine. Whine. Whine.

Smog is always a potentially serious problem in late July through the middle of August in the Los Angeles Basin, which is when and where the Olympic Games were held, because a strong upper-level high pressure system typically resides over the region, resulting in little air flow, stagnant air masses, and plenty of smog. A little gift from the Greek god Olympus (it *was* the Olympics after all), however, alleviated some of the smog problems—at least during the start of the Olympics in 1984.

A weak upper-level storm system was located over the eastern Pacific. The storm wasn't significant enough to bring any rain, but the presence of the weak low pressure system was enough to stir the atmosphere, resulting in less smog than usual during the first week of the Olympics. The typical summer upper-level high pressure system returned during the second week of the games, but since it generally takes a few days for the smog to return, it never became as bad as it normally is.

Snowplow Bowl

Snowstorms in New England during the winter are as common as controversial football games played in the snow in New

Most Polluted Cities in the United States

1. Los Angeles/Long Beach/Riverside, California
2. Visalia/Porterville, California
3. Bakersfield, California
4. Fresno/Madera, California
5. Pittsburgh/New Castle, Pennsylvania
6. Detroit/Warren/Flint, Michigan
7. Atlanta/Sandy Springs/Gainsville, Georgia
8. Cleveland/Akron/Elyria, Ohio
9. Hanford/Corcoran, California
10. Birmingham/Hoover/Cullman, Alabama

England, and one of the most controversial occurred on December 12, 1982.

A storm system moving northward through the Atlantic Ocean resulted in a raging snowstorm in Foxboro, Massachusetts—the home of the New England Patriots—while the Patriots played the Miami Dolphins. The bitterly cold, snowy conditions contributed to a scoreless game well into the final quarter. With New England moving into position to kick the winning field goal late in the game, and with snow covering the field, New England coach Ron Meyer ordered the grounds crew to use a snowplow to clear a spot in the field for his kicker. It worked. John Smith, taking advantage of the improved traction, scored the only 3 points of the game with his field goal.

Using the grounds crew to gain an unfair advantage should have been against the rules, but since the league had never considered this particular possibility before, it wasn't. The kick was permitted to stand, giving New England the win. After the season, the league prohibited the use of snowplows on the field during a game.

This Team Can't Win in the Cold

When winter arrives, so does the belief that football teams (professional and college) from the warmer southern climates can't win in the cold. This is said nearly every time two climatically uneven teams play each other on the field of the colder team.

The argument is based more on assumption than on analysis. Although there might be some logic to the statement if the

Tuck Rule Game

The storm system that produced a round of heavy snow during a playoff game between the New England Patriots and the Oakland Raiders on January 19, 2002, was not as intense as the storm that produced snow two decades earlier (see "Snowplow Bowl" earlier in this chapter); however, the controversy flew just as intensely as the snowflakes.

With a wet, cold, and slippery ball from the burst of heavy snow, quarterback Tom Brady made what seemed to be a potentially game-costing fumble; however, the official ruled that his arm was moving forward when he lost control of the ball, so it was ruled an incomplete pass rather than a fumble. This is called the tuck rule. Not only did the Patriots retain possession of the ball, kick a field goal, and win the game, but they went on to win the Super Bowl.

team from the northern states were composed of more players accustomed to the cold, since there would be some advantage to familiarity of playing in low temperatures, this is generally not the case, especially in the NFL. A typical NFL roster is composed of players from all over the country, not just from near the city hosting the team. Besides, if playing in cold weather were such a prerequisite for a good performance, then the southern team would most likely rent outdoor practice facilities in Saskatchewan in order to prepare for the big game.

In a contest between two evenly matched teams, the home team is most likely to win in the cold for one reason—the home team is most likely to win in general. The record of a team from a warm climate playing in a cold city likely varies little from the record of any team playing on the road.

6

What Does "Red Sky at Night" Mean?

WEATHER WIVES' TALES

THIS MAY BE HARD TO believe, but once upon a time, in the distant, distant past, weather forecasts were not perfect. Sometimes a meteorologist would forecast sunshine, but it would rain; sometimes he would predict rain, but the sun would continue to shine. Thank goodness those days are behind us.

Okay, okay, you can stop laughing. Weather forecasting is not perfect, but it's much better than it was before there were weather balloons, satellites, radar, computer models, four-year bachelor of science degrees, and 10-year student loan bills. Before those things, observation and experience were the only basis for prediction, which were often captured and recorded in the form of old wives' tales and adages.

Viewing these mostly antiquated phrases through the eyes of science shows how our ancestors were often fairly decent non-scientific weather forecasters. Some of the phrases, though, make me think that the apple cider fermented a little too long—if you know what I mean.

Cold Is the Night When Stars Shine Bright

Forecasting the weather by the brightness of the stars, as is the case with the adage that "Cold is the night when stars shine bright," is a difficult chore, not because of an inherent inaccuracy of the statement but because of air and light pollution. In 21st-century America, locations where the stars are visible enough at night to make such a forecast are limited.

Residents of the 13 original colonies based reasonably accurate short-term forecasts on the adage, though. Brightly lighted stars are an indication of two things: a cloudless sky and a dry air mass, which are the conditions needed for maximum cooling to take place during the course of a night.

Clouds act as a blanket of sorts, preventing the heat that collected during the course of the day from radiating away from the ground. A cloudy night will not be nearly as cold as a cloudless night with all other meteorological conditions being equal. Not all clear nights are equal for stargazing, though, because on many clear nights, the stars will be faint. The faintness can be caused by clouds too thin to block the starlight completely, but it's more likely caused by haze resulting from moisture or pollution in the air. Brightly shining stars indicate little moisture (and no haze) in the air, so the night will become as cold as the air mass and length of the night will allow.

This adage will only be true, of course, when clouds do not arrive during the course of the night, which is generally the case when the air mass is clear and dry to start. Even when the forecast is accurate, it has limited usefulness, since it doesn't give an indication of how cold it will be.

Cows Lie Down When Rain Is on the Way

Animals are much more in tune with their environments than are human beings, which is why few barns are equipped with high-speed Internet in order to get the latest forecast information. Many a farmer, though, has observed that "Cows lie down when rain is on the way."

I've always had a tendency to think that cows lie down when they're tired, not when it's going to rain, but it makes sense that cows, especially modern cows bred in a way to promote the largest size possible (can you say steroids?), would have knee joints affected by moisture. The knee is often a problem for professional human athletes (can you say steroids?) because the ligaments and tendons around the knee do not get larger and stronger as muscle in the rest of the body does, with the result being a large body with regular size knee tendons and ligaments.

Just as moisture often affects *our* joints, it would affect a cow's, prompting a cow to lie down when in discomfort. The potential problem with the statement, of course, is that moisture in the air does not automatically mean that rain is on the way. However, increased humidity followed by rain might happen

Animal Lore

"Cows are more likely to graze in the lowlands before a storm." This might be true because of joint discomfort due to the lowering pressure (who wants to walk when her legs hurt?) or to avoid the wind often associated with an approaching storm. Either that, or the cows don't want to bother climbing the hill if they're just going to lie down anyway.

"When a sea gull sits on the sand, it's never good weather while you're on land." Whether it's the wind associated with an upcoming storm or the roughness of the seas, it is not surprising that gulls take to land when a storm is coming.

"When goats seek shelter, expect a storm." Be very frightened when they ring the doorbell. Actually, evidence of goats seeking shelter before a storm is lacking; however, with their wool coats, they do seek shade during sunny, hot days to keep from becoming overheated. Because many summer thunderstorms occur on partly sunny, hot days, this might help explain the tale. Perhaps they're not seeking shelter from the upcoming rain but, rather, from the heat on a day when a thunderstorm is likely to follow.

frequently enough for the wives' tale to have once been a mildly effective forecasting tool.

If You See the Backs of the Leaves, It Will Rain Soon

Wind direction is an important part of any weather forecast and the basis for some of the popular weather adages, including "If you see the backs of the leaves, it will rain soon."

Some form of an easterly wind (wind from the east, northeast, or southeast) is needed in order for the backs (or undersides) of a tree's leaves to show, and since an easterly wind often means that rain is on the way in parts of the country, the statement has validity. In other parts of the country, though, it's as wrong as can be.

When the wind becomes easterly, it often means that a low pressure system is approaching from the west. The wind around a low pressure system blows in a counterclockwise direction; put a low to your west, and the wind will be easterly. Because low pressure systems typically track from west to east across the country and often accompany rain (or snow, but that's typically when the trees are without leaves), seeing the backs of the leaves might, indeed, indicate that rain will soon follow. This works reasonably well for most locations to the east of the Rockies, which is where most early Americans resided (when the phrase was most likely to have been used as a forecast).

The east wind, though, can also occur to the south of a high pressure system's clockwise flow. This easterly flow might be enough to produce rain along the Eastern Seaboard, since an easterly wind, even one not associated with a storm, might produce rain and drizzle from the Atlantic. Elsewhere, though, a high pressure system to the north typically produces fair weather.

Along the West Coast, an easterly wind not associated with a storm system is usually a dry wind because it's originating from land, and a high pressure system over interior parts of the West often results in a strong, very dry wind, especially in California. Wind speeds can exceed 100 mph, so it's more likely that they'll blow the trees over than innocently turn the leaves over.

In Like a Lion, Out Like a Lamb

To a meteorologist, the months are like children, meaning the weather of each is equally special and lovable, and he would never pick a favorite. A parent might single out the child who's most likely to cause the most trouble, though, and for a meteorologist, March is that problem child. In one moment, he's lovable and sweet, and in the next, you know confusion at the hospital means you ended up with Lucifer's kid. The commonly used weather phrase about the month of March, "In like a lion, out like a lamb," is an indication of March's weather variability, but the phrase has been used in a couple of ways.

The erratic nature of March is what many people think of when they say, "In like a lion, out like a lamb" and add the inverse, "In like a lamb, out like a lion" to complete the adage. This extended phrase is used to signify that March weather is changeable (notice I didn't say "unforecastable!"), and that if the weather is stormy to start the month, then it will be peaceful to end, but if the weather is peaceful to start, then it will be stormy at the end. Either way, you get some of both.

The shortened version of the phrase, simply "In like a lion, out like a lamb," is used to indicate that March is the meteorological transition from the harshness of winter at the start of the month to the more peaceful days of spring at the end of the month. While that's good in theory, I've always found that March was too fickle for this usage (see "March Extremes" later in the chapter).

This phrase might actually have nothing to do with the weather at all; instead, astronomers lay claim to the phrase because of the position of two animal-shaped constellations during March. Leo the Lion is visible at the beginning of the month, and Aries the Ram (which supposedly represents the lamb) is visible at the end of the month.

Which way have you been using the phrase all of these years?

It's the Calm Before the Storm

Many of the entries in this chapter have not been actively used for decades, but we all have said "It's the calm before the storm" in any number of ways in our daily conversations. Most of this usage is not related to the weather. I hope the accuracy of the statement is greater when used in the non-weather realm.

"It's the calm before the storm" is typically uttered when the speaker knows that a storm is expected to follow. Calm weather is followed by more calm weather, or at least mildly unsettled weather, too often for this to not be the case. Coastal Florida residents might comment that it's the calm before the storm the day before a hurricane arrives because they know the storm is approaching. Residents in New England mention the calm before the storm the day before a Nor'easter arrives because a blizzard watch is in effect. California residents say it's the calm before the storm the day before heavy rain is forecast because the Pineapple Express (see the box later in this chapter) storm was the lead story on the evening news. Without this forecast information, everyone would go about his or her day without knowing that the calm weather was about to be followed by a storm.

March Extremes

March extremes are so common that the month almost always brings wild changes. I selected March 2007 for this example not because it was exceptionally erratic but because it wasn't. These changes represent a typical wild March across the United States.

New York City	**LAMB**: 67°F on the 15th	**LION**: 6 inches of snow, 36°F on the 16th
Chicago	**LAMB**: high of 78°F on the 25th	**LION**: low of 15°F on the 6th
Los Angeles	**LION**: high of 67°F on the 9th	**LAMB** (a warm lamb): high of 92°F on the 11th
Houston	**LION**: cool with 6 inches of rain on the 12th–14th	**LAMB**: 10°F warmer than normal at the end of the month

Meteorologically speaking, the instances when this is a valid statement are extremely rare, occurring too infrequently for the phrase to be a reliable predictor of future weather. When a powerful thunderstorm is approaching, the intense updraft of the thunderstorm occasionally counteracts the normal wind in advance of the thunderstorm, resulting in a breezy day becoming calm before the storm arrives.

In those circumstances, the upcoming stormy weather creates the calm; in the others, the knowledge of an impending

Phoenix	LAMB (nearly ready to come out of the oven, I might add): high of 99°F on the 15th	LION: high of 66°F on the 28th
Seattle	LION: snow on the 1st–2nd	LAMB: 68°F on the 6th
Miami	LAMB: high of 90°F on the 3rd	LION: high of 79°F on the 31st
Atlanta	LION: high of 48°F on the 4th	LAMB: high of 88°F on the 25th
Denver	LAMB: high of 75°F on the 18th	LION: high of 41°F with 4 inches of snow on the 29th
Minneapolis	LION: highs of 23°F on the 3rd and 49°F on the 30th	LAMB: high of 81°F on the 26th

storm prompts the statement while the weather is still relatively calm.

Long-Range Lore

While the advantages of knowing what the weather will be in 4 minutes, 4 hours, or 4 days are obvious, the interest in the weather in 4, 5, or even 6 months into the future has tremendous

Pineapple Express

While the term *Pineapple Express* might sound like a Dole truck exceeding the posted speed limit, it's actually a weather term commonly used—and misused—in the western part of the United States. *Pineapple Express* is intended to describe a storm, or more likely a series of storms, containing tropical moisture from near Hawaii as it arrives along the West Coast.

The term is often misused, though. Not all West Coast storms bearing tropical moisture originate near Hawaii. Moisture is sometimes pulled northward from the tropical regions much closer to California in an area where there are no islands—and therefore no pineapples. While it's not technically the Pineapple Express, it's become common for the media and residents to blame the Pineapple Express any time it rains heavily in the West, which either oversimplifies the weather or complicates the pineapple-growing industry.

advantages as well. We demand accurate long-range forecasts more than we used to, but we probably needed them more in the past. In colonial America, for instance, the weather for the next growing season often meant the difference between plentiful food and scarcity, and the weather for the next winter often meant the difference between a winter of health and a winter of desperation and illness.

Long-range weather proverbs are typically an indication of the natural variability of the weather, not legitimate forecast information. Heat waves are always followed by cold snaps, droughts are always followed by plentiful rain, and wet periods are always followed by dry periods; if they weren't, then the weather would always be moving toward a certain end in an unstoppable march. Using the natural variability of weather as a forecast, though, is not effective because you don't know when this inevitable change is going to happen.

No Weather Is Ill If Wind Be Still

The weather wives' tale that "No weather is ill if wind be still" has a certain charm to it, most notably the use of the word *ill* to

Long-Range Proverbs

"If autumn leaves are slow to fall, prepare for a cold winter." If trees are an accurate indicator of future weather, then I would expect the opposite. A cold winter would most likely mean the potential for early snow, and trees are heavily damaged when snow falls on a tree with leaves. If long-lasting leaves were a precursor of anything, it would most likely be of a late-arriving winter; otherwise, trees would be predicting their own demise. This proverb is most likely an indication of the belief that the weather is likely to change, and a long, warm fall with lingering leaves is most likely to be followed by a long, intense winter.

"A windy winter, a rainy spring." Winter storms might spill into spring, and a windy winter might be an indication of a more-active-than-normal storm track, so this passes the meteorological commonsense test; however, forecasting is more complicated than that. Trust me.

"A warm summer means a cold winter; a dry spring means ample summer rainfall." Forecasting the opposite of what's going on now will always eventually be correct. It's the same with "warm October, cold February." The law of averages is not a forecast.

describe the weather. I'm going to start using that line: "Ill weather today; don't forget your umbrella."

A lack of wind followed by healthy—I mean, fair—weather was observed with enough regularity for early weather observers to assume that a lack of wind meant a lack of precipitation. The statement, though, would be as simple as assuming that a strong wind means rain is on the way, which is also an oversimplification of the weather.

Now, if we were to add "and the sky is clear" to "No weather is ill if wind be still," then there's an excellent chance that we'd have a fairly accurate short-term forecasting tool because fair weather will continue for at least a while longer, since a clear sky and a lack of wind most likely means that a fair weather high pressure system is over the region. Perhaps you could say, "If the wind be still and the sky clear, then the weather you do not have to fear."

If the sky is not clear, however, then it might mean that precipitation is on the way despite the lack of wind. When a large storm system is approaching, even if it's a storm with a lot of wind, clouds often precede the storm by hundreds of miles, perhaps even extending into a high pressure system. The clouds are often capable of producing precipitation long before the wind associated with the storm arrives, meaning that precipitation might soon follow despite a lack of wind.

It's worth noting that the logic of this statement contradicts the familiar calm before the storm adage (see "It's the Calm Before the Storm" earlier in this chapter). Calm weather means either that a storm is on the way or that dry weather is on the way—but you can't have it both ways.

Rain Before 7, Fine by 11

Most weather proverbs are based on logic—even if incomplete logic—but I think whoever created the line "Rain before 7, fine by 11" was just a frustrated poet. The adage applies to the morning (7:00 to 11:00 a.m.), because what happened at night was of less importance in the dusk-to-dawn world of our ancestors.

A certain weather event is more likely to create rain overnight, spilling into the morning hours before ending—called a Mesoscale Convective Complex (MCC). When a southerly flow of warm and moist air clashes with a cooler, northwesterly flow, thunderstorms rage during the overnight hours. These nocturnal (nighttime) thunderstorms occur with the greatest frequency in the Plains and upper Midwest, with the dying remnants (usually just a band of light or moderate rain) moving through the East or Middle Atlantic region the next morning. This is, no doubt, the source of the saying.

Most rain-bearing storms, however, are as likely to produce rain in the late morning and afternoon as the early morning. This might help explain why weather poetry has never taken off.

A Rainbow in the Morning Is the Shepherd's Warning; a Rainbow at Night Is the Shepherd's Delight

What must early man have thought when he saw a rainbow—an array of bright colors arching across the sky—while, rain was falling and the sun was shining. This magical display is why he gave credit to the gods or came up with the idea that there must be a pot of gold at the end of the rainbow. The magic is probably also why early weather observers made forecasts based on rainbows.

People tell me that the sun rises in the east (I try to avoid waking up that early), and because rainbows occur on the opposite side of the sky as the sun, a morning rainbow is an indication of a shower to the west. Since the weather generally moves from west to east across our continent, the assumption is that the shower to the west is headed our way, and thus "A rainbow in the morning is the shepherd's warning" (also stated in the sheep-free version as "A rainbow in the morning gives you fair warning") is logical.

Similarly, with the sun setting in the west (I can personally attest to the accuracy of that one), an evening rainbow is an indication of a shower to the east. The assumption that the shower has already passed the local area is also logical and thus "A rainbow at night is the shepherd's delight." (It would be more accurate to say evening, not night, since rainbows can't occur after dark, but who am I to argue with an ancient poet?)

If one lonely, little shower is occurring, then this simple forecasting method will work well. The evening shower to the east will most likely, but not always, remain to the east, and the morning shower to the west has a reasonable chance of bringing rain to the observer. Give a gold weather vane to the poet.

A shower occurring while the sun is shining, though, is an indication of an unstable atmosphere, with an excellent chance that more than one shower is in the area. Even if the rainbow-producing shower remains to the east, additional showers might arrive from the west, so a rainbow in the evening could mean that the shepherd should keep his wool fleece nearby, just in case.

Red at Night, Sailors Delight; Red in the Morning, Sailors Take Warning

If I were a sailor, I'd want all of the latest weather information. I'm getting on the boat with all of the latest satellite data, radar data, and computer forecast maps, not the one stuck with basing a forecast on the color of the sky, such as "Red at night, sailors delight; red in the morning, sailors take warning."

This statement is like fishing with a net that has a hole in it; you might catch some of the big fish, but the little ones will get away. In the mid-latitudes of the Northern Hemisphere, the weather generally moves from west to east—the basis for the statement. A sunrise or sunset is more red in color when the atmosphere contains more moisture and dust, which are both indicators of a storm system. The moisture part is obvious, and the dust part is a result of wind typically associated with a storm; thus, a red sunset must mean a storm is moving away because the moisture and dust are to the east. If the sunrise is red, then a storm is approaching because there is moisture and dust to the west.

The wives' tale might forecast large-scale storms from a normal direction, but it could fail when the humidity is high enough to result in a red sunrise without a storm. The adage's biggest failing, though, would be for small-scale storms, such as an individual thunderstorm or a small cluster of thunderstorms, because these can occur on a scale too small to affect the color of the sunrise.

See why I am getting on the boat with modern equipment?

Rain Foretold, Long Last; Short Notice, Soon to Pass

Weather folklore is often too simple to be of practical use today, but imagine the utility of such a phrase before scientific study, mass communication, and weather tools came into being. It can give us a level of appreciation of the insightful observational skills of the time. If you were a farmer, hunter, or fisherman, then knowing "Rain foretold, long last" and "Short notice, soon to pass" could certainly help you plan your day before the Weather Channel came along.

Generally speaking, the type of precipitation event that seemingly comes out of nowhere, such as a spring shower, summer thunderstorm, or winter snow squall, does not last long (short notice, soon to pass). These are often the result of a convective weather pattern (see "Convection—and Not in My Oven" in Chapter 1), the nature of which generates multiple precipitation cells intermixed with a partly cloudy or a clear sky. One minute, it's sunny, and the next, snow or rain is falling from the sky, perhaps intensely. Then, within minutes, the sun might be shining again.

On the other hand, long-lasting rain precipitation events—a spring soaking rain, winter rainstorm, or major snowstorm—are typically preceded by a gradual increase in the amount of cloudiness and a lowering of the cloud base. This process, which might take 12 or more hours, often leads to steady precipitation that might also last for 12 or more hours (rain foretold, long to last).

The adage does not give the level of detail a modern audience might demand, such as the type of precipitation or the intensity, but these words helped many an early outdoorsman plan his daily tasks, giving some guidance about how long the precipitation would last. You might say: "An accurate guess, less time in the mess."

Ring Around the Moon, Rain Will Come Soon

When weather forecasting was as accurate as your average weather wives' tale, one of the most reliable methods was to look at the clouds. We make predications based on specific cloud types all of the time. For example, when you're at the family reunion and have a 20-minute walk from the car to the pavilion, the towering clouds quickly closing in means a forecast of soggy potato salad. This type of sky watching, minus the potato salad, is the basis for a forecast of rain when there's a ring around the moon.

The clouds aren't as clearly visible at night, of course, but when there is a thin veil of high clouds covering the moon—typically cirrus clouds—the moon sometimes appears as if it were surrounded by a ring of light. This "lunar halo" is caused by refraction of light through the icy crystals composing the high clouds, often 20,000 to 30,000 feet above the ground.

In most instances, a storm bearing steady rain is preceded

by a sky full of cirrus clouds, so it's not surprising that this phenomenon was witnessed often enough for early weather observers to deduce that a ring around the moon was to be followed by rain. The cirrus clouds can precede the rain by as little as a couple of hours or by more than 12 hours. That might not be enough weather detail for today's weather consumer, but for those who predated the *Farmer's Almanac*, it was a winning forecast.

Even these early meteorologists had their forecasting problems, though, since cirrus clouds can also precede a *snow*-producing storm, which might fall into the realm of a reasonable wives' tale forecast. Cirrus clouds, however, are often produced by something other than a rain-bearing storm, so the ring around the moon prediction could be inaccurate. Cirrus clouds can be the remnants of a dying thunderstorm—one no longer producing rain—or can simply be the result of a thin layer of moisture in the upper levels of the atmosphere, not associated with a storm on the way.

In the modern world, the remnants of jet contrails can even produce a thin layer of clouds that results in a ring around the moon, or at least a partial ring. In that case, a more correct phrase would be: "Ring around the moon, lost luggage soon."

Summer Fog for Fair, Winter Fog for Rain; a Fact Most Anywhere, a Valley or a Plain

No one likes a meteorologist with hubris because his reputation is only as good as tomorrow's forecast, so I wonder how people felt about the inventor of this overly confident weather fog lore: "Summer fog for fair, winter fog for rain; a fact most anywhere, a valley or a plain." Who does he think he is—or, more accurately, who did he think he was?

Different fog types form in different ways (see "In a Fog" in Chapter 1), and one of the most common types is radiation fog, which, by the way, has nothing to do with an accident at the nuclear reactor. This radiation refers to the escape of the daytime heat into the night, allowing the air to cool to a point of condensation, creating fog. This occurs most effectively when the night is clear, meaning the weather is fair. The adage refers to "summer fog for fair"; however, while fog sometimes forms in the summer, the nights are generally too short for this to occur with as much regularity as it does in the fall. This type of fog is also much more common in a valley than on a plain or the mountains—as opposed to "a fact most anywhere."

Summer fog tends to be more short-lived and follow a heavy shower or thunderstorm, when the air has been cooled, fresh moisture is on the ground, and the air has calmed. This can be a radiation fog, and it might also be what we call a steam fog (fog produced by the cooling of a warm surface, just as dumping water on a hot pan produces steam). Regardless of its classification, it often indicates fair weather is coming because summer showers and thunderstorms are often isolated (one per customer, please); however, that's not always the case.

Winter and spring fog in much of the country is more likely to be associated with an ongoing rain event, which applies to the "winter fog for rain" part of the saying. Much of winter and spring fog is a result of the high humidity associated with falling rain, especially if it's a mild rain over cool or snow-covered ground. This is a fact almost anywhere—in the valley, plain, or mountains.

In the western part of the country, there is a major exception to both counts—the "winter fog for rain" and the "fact most any-where, a valley or a plain." The valleys of the West often fill with dense fog during the winter, and it's the fair-weather radiation type of fog discussed earlier.

Perhaps our fog expert should have traveled more before making such bold proclamations.

When Windows Won't Open and Salt Clogs the Shaker, the Weather Will Favor the Umbrella Maker

When I was struggling through calculus, physics, and thermo-dynamics in college, I had no idea that early weather forecast-ers studied windows, salt shakers, and umbrella makers. If those were the required courses, then I'd love to have seen their electives.

The adage "When windows won't open and salt clogs the shaker, the weather will favor the umbrella maker" is about mois-ture in the air. All window frames used to be made of untreated wood, which expanded when exposed to moisture, making them difficult to open when the air was thick with moisture. Salt ex-pands when it collects moisture from the air, clogging the shaker. To me, by the way, nothing says fine dining like a salt shaker with a few grains of moisture-absorbing rice mixed in.

Anyway, the wives' tale is based on the premise that heavy

Vog—Fog with a V

Vog, with a *V*, is a combination of fog and volcanic ash, which occurs in Hawaii. Vog can occur when trade winds weaken and an upper-level high pressure system is bringing tranquil weather, allowing for a stagnant air mass. Oh, and there has to be a percolating volcano nearby.

moisture in the air—enough to stick windows and clog salt shakers—should be enough to result in rain soon. I understand the logic; however, it's too simplified to be of any practical use. A dense fog means that the air is thick with moisture, but it doesn't guarantee rain. It's the same with high humidity in the summer; it doesn't necessarily portend rain.

I'll stick to forecasting rain the new-fashioned way, and I'd like to ask restaurant owners who use the rice trick a question: How many years has it been since you cleaned the salt shakers?

When Dew Is on the Grass, Rain Will Not Come to Pass

Because I live in a housing complex with a common area (shared by all residents), the first thing I think of when my shoes get wet while stepping onto the grass is that I hope that it's dew—and not something worse. After the relief of realizing it wasn't from a neighbor's dog, the weather rhyme "When dew is on the grass, rain will not come to pass" might wander into my mind, since it's generally accurate.

Some Salt-Shaker-Clogged Cities

Most Humid Cities (Based on Relative Humidity, Not Dew Point):

Forks, Washington	83.0%
Olympia, Washington	78.0%
Port Arthur, Texas	77.5%
Lake Charles, Louisiana	77.0%

When the cool air of a clear night condenses into water droplets, we call it dew (which is why we call a unit measuring humidity the dew point). Notice the part about a clear sky and dew formation, and this clear sky is the basis of the "rain will not come to pass" part of the rhyme.

A clear, cool night is followed by dry weather often enough for this phrase to have had some usefulness for our ancestors, especially if the dew formed early enough in the night for them to forecast a nice day for the next morning.

When the Wind Is Out of the East, the Weather Is Not Fit for Man or Beast

The basis for the saying "When the wind is out of the east, the weather is not fit for man or beast" is the same as that for "If you see the backs of the leaves, it will rain soon" (discussed earlier in this chapter). The former adage is just more dramatic.

If you'll recall from the earlier entry, a wind from the east often means that a low pressure system is approaching, so especially from the Rockies to the Atlantic Coast, this east wind often means the impending arrival of rain. A low pressure system, or storm, does not always mean an intense weather event, of course—certainly not one that is guaranteed to be too difficult for a *beast* to bear, because a beast is a little more weather tolerant than your average person. The intensity aspect makes me believe that this phrase most likely originated along the East Coast, probably in New England.

The speed of wind is greatly affected by friction. The more ground, trees, buildings, mountains, and, I guess, *beasts* that interrupt a wind, the more the speed of the wind decreases. Wind that blows directly over water is exposed to much less friction than air that blows over land, so along the Eastern Seaboard, an east wind produced by a storm is often stronger than an east wind produced by a storm in a land-locked area.

More important to this myth is that the warm water of the Atlantic fuels winter storms, so these wind-bearing storms are often much more intense and much windier than their counterparts in other areas of the country. That's why an east wind has the reputation of being intimidating for both *man* (probably Puritan man) and *beast*.

A Year of Snow, a Year of Plenty

I've never met a happy farmer. No matter what the weather is, farmers want something different. When it's dry, they want rain—but not too much. When it's wet, they want dry weather—but not for too long. When it's hot, they're afraid that it's going to be hot for too long, and when it's cold—oh, don't even get them started about the cold. That's why if they want to believe the optimistic and simplistic wives' tale "A year of snow, a year of plenty," I don't argue with them.

Meteorology provides a basis for the statement, but for any long-range prognostication, including those by professional meteorologists in the 21st century, accuracy is limited. A couple of advantages exist for farmers when the ground is wet with the melting winter snow. The wet ground means that the soil will stay moist longer into the spring, providing a good start to the growing season, at least as long as it doesn't stay too wet for too long (or rain too much in the spring, keeping farmers from planting). In other words, if it does stay too wet for too long, a more accurate adage might be "A year of snow followed by spring rain, a year of root rot."

Nor'easter or No'nor'easter?

While it has become fashionable to call nearly every storm to affect the eastern part of the country a Nor'easter, it is not accurate. A Nor'easter is a storm that develops along the Carolina or Middle Atlantic coast and then strengthens as it moves northward. It's called a Nor'easter because the winds preceding the passage of the storm are from the northeast. In a sense, it's a storm that can occur during any time of the year; the classic Nor'easter occurs during the winter or spring but also occasionally during the fall.

Snow in the Breadbasket

The snowiest winters for a few locations in the Plains and Midwest—it looks as if 1912 must have been a year of plenty.

Sioux Falls, South Dakota	94.7 inches	1968–1969
Omaha, Nebraska	67.5 inches	1911–1912
Springfield, Missouri	54.5 inches	1911–1912
Wichita, Kansas	39.7 inches	1911–1912
Tulsa, Oklahoma	25.6 inches	1923–1924

The other advantage is a slight decrease in the danger of a spring frost because the higher moisture in the ground means that the air will have a little more moisture in it, and moist air does not cool as well as dry air does. I know, this is a bit of a stretch, but if it helps reduce medical bills related to stress, then it's good enough for me.

"A year of snow, a year of plenty" doesn't apply directly to western farms, where many of our crops are grown, but it does apply indirectly. Summers are always very dry in the West, even in the Pacific Northwest (see "It Always Rains in Seattle" in Chapter 2), so plentiful snow in the mountains, not on the farm itself, is needed for a year of plenty because summer irrigation depends on melted winter snow.

That's not a wives' tale; that's a fact of farming in the dry West.

Planes, Trains, and Crops

HOW WEATHER AFFECTS TRANSPORTATION AND CROPS

WHILE TRANSPORTATION AND CROPS MIGHT seem like strange companions for a chapter, the weather can lead to joy or tragedy (or feast or famine) for both.

The dangers of bad weather for transportation are obvious in many cases, but even a relatively innocent-sounding snow shower can result in massive traffic accidents. Even modes of transportation generally seen as impervious to weather, such as trains, can be negatively affected, and most of us know that the weather can affect whether our car will operate properly.

As I mentioned in the last chapter, I have never spoken to a happy farmer, but is that any surprise? Extreme weather events, such as a major freeze or a hurricane, can cause financial hardship for years. Similarly, relatively minor weather events, such as an ill-timed frost, a couple of weeks of dry weather, or a thunderstorm with hail, can make the difference between a success and a failure of a given crop in a given year. Even overly bountiful years can conceivably have a negative impact on the amount of money earned from crops by driving prices down.

My farming knowledge is from my experience as a meteorologist, not as a firsthand farmer; otherwise, this chapter would be very short. My limited farming experience includes time on my grandparents' farm on summer vacation when I was young, when I tried to find a way to sleep until 11:00 a.m. (which is about 5:00 p.m. farmer time).

As a meteorologist, I have learned to help farmers from behind my computer rather than from behind the wheel of a John Deere. Believe me, we're all better off this way.

Air Temperature and Tire Pressure

You don't need a meteorologist to tell you that proper tire inflation is important for safe travel and for the longevity of a tire, but you might benefit from a meteorologist's knowledge about how air temperature affects tire pressure.

Because hot air is lighter and more buoyant than cold air, air molecules move more quickly and spread out more, so when in a confined area (such as the inside of a tire), hot air will create a higher pressure than will cold air. In other words, there doesn't have to be a change in the *amount* of air inside of a tire for the *pressure* to change; a change in air temperature will result in a change of tire pressure—and this change is sometimes dramatic.

When the air temperature is extremely cold (zero or lower, for example), checking the tire pressure while the car is outside will result in an inaccurately low tire pressure reading. Filling the tire to meet the vehicle's recommended levels while the tires are extremely cold will result in over-inflated tires once they warm while driving. Some estimates are that the tire pressure loses 1 pound for 10°F, so at zero, the tire pressure could be 5 or 6 pounds lower than when the tires are warm. This is affected by the size of the tire, though. On a larger tire, such as on a pickup or delivery truck, it might be less, but for a small car (with smaller tires), it could be even more.

Assault and Battery

It's happened to all of us. When we're running to the store for the Twinkies and nachos before the big game or going to work for the important meeting, we turn the ignition and all we hear is a click—the battery is dead.

We should expect that to happen every 4 years or so unless we're smart enough to replace the battery before it dies. (Be honest; don't you hate people who plan ahead?) Those of us who live in the colder climates are more susceptible to being stranded on the coldest morning of the year because we all know that cold weather is tougher on batteries than hot weather, right? No, extreme heat is, in fact, more likely to cause battery failure.

Extreme heat causes batteries to corrode, and this heat-related corrosion is the leading cause of battery failure. In fact, many of the failed batteries of the winter fail because they were weakened by summer heat, and in these cases, winter cold gets the assault and battery charge.

To be fair, extreme cold is very hard on the life of a battery. Not only is the car hard to start because the oil is heavy and thick, but we also tend to take more short trips, which might not allow the battery to recharge completely. Protecting the battery against extreme cold is easier than protecting it from extreme heat. Many times in bitterly cold climates, the question, Did you remember to plug the car in? is asked before the lights are turned off, not because most people own an electric car but because the device designed to keep the engine or battery warm over-night needs to be plugged in. Some people cover the battery with a blanket at night—kind of like a car tuck-in—but batteries aren't like people: They don't produce heat that can be con-tained by a blanket. A blanket will only keep the heat of the warm engine from dispersing after it's been turned off, so it has limited usefulness.

Some of us start the engine a couple of times per night when it's bitterly cold in order to keep the battery charged and fresh. If you don't let the car run long enough to recharge the battery, however, then you're putting on your long underwear, three layers, and that ugly skullcap only to make it more likely that your battery won't start in the morning.

Beans in the Teens

Although it's illegal to gamble on sporting events in much of the country, there is still a high-risk gambling venture that's available—futures trading. This type of gambling is often driven by long-term weather forecasts, which is why I'm choosing to mention it here.

In a very basic sense, a futures investment is based on the concept that prices for commodities can drastically change, and those who are providing the commodity (a farmer with a soybean crop, for example) might be willing to lock his price in at the given market value during the beginning of the growing season rather than run the risk that prices will be much lower by the time the season is over. In other words, if the price drops over the season, he will make more money than the farmer who sells at market value at the end of the season. Of course, if the price rises, the farmer will make less money than his neighbors. It's a calculated risk.

For the investor, who accepts the responsibility for paying the difference between market value now and the change in market value later, the risk is *not* calculated and can be extreme—as is the potential reward. If the prices rise significantly, he receives the additional profit the farmer would have made had the farmer not decided to lock in his price. If the prices fall significantly, though, the investor is responsible for the difference between those prices, which can exceed the initial value of his investment. In other words, he can lose more than his original investment—he can lose money he didn't even invest in the first place. Futures investing is much more complicated than what I know and can discuss here, so don't make any decisions based on what I'm saying, especially before reading (think Paul Harvey) the rest of the story.

A certain meteorologist who shall remain nameless (but he

has gone on to write books on various topics) once heard a radio commercial talking about a futures investment. The announcer talked about flooding in the Midwest and how that would drive soybean prices upward ("Beans in the Teens" was the theme of the commercial). He, being a meteorologist, knew of ongoing Midwest flooding and knew that it would likely get worse before getting better, so he invested a small amount of money (very small, fortunately) in soybean futures.

The rain fell and flooding got worse, but as the summer progressed, soybean prices fell and fell and fell—into the single digits—instead of rising through the teens as expected. While the flooding caused damage to some soybean fields, an abundance of rain without flooding was widespread enough to produce a bumper crop of soybeans overall, driving the price down.

Control the Soybean Market, and You Control the World

Okay, controlling the world might not be as simple as controlling the soybean market, but so many products are made of soy (vegetarian meat products, dairy substitutes, candles, and even plastic) that it might not be far from the truth. The United States is in a good position for soy world dominance because it's the world's leading soybean producer. In the year 2000, 72.7 million acres were devoted to growing soybeans.

This magic bean requires about the same conditions as the average green bean grown at home, except that it needs a longer period (3 to 5 months) of frost-free weather and prefers slightly higher temperatures.

This meteorologist, who is very, very close to me, has not gambled on the weather since; however, futures investments based on weather forecasts is a huge part of the investment world, and many meteorologists are employed by those who are looking to make such investments.

Black Ice

Many of us consider *ice* to be a four-letter weather word—not because we can't count or spell but because ice is dangerous. The danger of most types of frozen precipitation (see "Precipitation Types" in Chapter 1), such as hail, sleet, and freezing rain (I'm glad that I'm not a mailman, by the way), is obvious, but the danger of a certain type of ice, referred to as *black ice*, is in its secretive nature.

Black ice is not actually ice that is black in color, as the name suggests, but a very thin layer of ice with no air embedded, so the ice is transparent. A roadway covered with a thin layer of ice appears to be its normal black color.

Black ice can form when there is a light mist, either falling mist (drizzle) or mist from dense fog. A driver might assume that the temperature is above freezing because the mist is in liquid form, but when either the temperature is below freezing or the roadway is colder than the air temperature because of recent cold weather, this mist will form an icy coating.

Not only is black ice invisible—and occurs when the weather is not overly stormy—but it can occur on a small scale or in an isolated area. The ice might be limited to a fog-filled valley or the top of a mountain; the roadway might be perfectly safe for hundreds of miles in either direction, so drivers might have no idea of the danger they're in.

Blinding Snow

Not many people react with terror when the forecaster says, "Mostly cloudy and cold with a snow shower today." Those who love snow aren't very happy because there won't be enough snow to bother with, and those who loathe snow aren't too upset because there won't be enough to cause problems. A snow shower—how bad could that be?

A quick, intense snow shower can, in fact, be one of the most dangerous types of weather, as evidenced by the many multi-vehicle auto accidents over the years, including two on I-80 in Pennsylvania within hours of each other on the afternoon of December 28, 2001. One was a 63-vehicle accident involving trucks with flammable liquids, resulting in a fire that burned for hours, and the second was a 50-vehicle accident a few hours later.

Both accidents were caused by short, intense snow showers (often called snow squalls), which pose multiple dangers. The nature of the event often results in unsuspecting motorists driving down the highway at 65 mph on a partly cloudy, cold day, who are suddenly—before they have time to slow down—transported into an intense snow squall or into the back of an existing multiple-vehicle accident caused by an intense snow squall. This double-accident day wasn't a one-time event; on the same roadway on January 6, 2004, three multiple-vehicle accidents occurred within 20 miles of each other, also caused by snow showers.

The intensity of the snow is, of course, a major contributor as well. Many a great snowstorm—one that produced a foot or two of snow—never had the intensity of a 20-minute snow shower, with its tiny inch of accumulation. The snow, which is often accompanied by a thunderstorm-like intensity of wind, results in an instant blizzard, with visibility reduced to a few feet, meaning that a driver cannot see beyond the front of his own car. Finally, the initial layer of this burst of snow often melts on the road surface (heated by cars or the warmth of the day) only to freeze into a thin layer of ice as additional snow (accompanied by falling temperatures normally associated with a burst of precipitation) falls on top of the now-icy layer.

In other words, unsuspecting drivers (meaning vehicles driving at, or above, the posted speed limit) drive into an instant blinding blizzard on ice- and snow-covered roads with visibility limited to their own hood.

An innocent snow shower, huh?

Bridge Freezes Before Road Surface

The Bridge Freezes Before Road Surface sign seems to be noticed only in July, but it is an important concept for drivers to understand.

Roadways freeze when the air is cold enough to allow snow or a collection of moisture (falling rain, melting snow, mist from a dense fog, or even a spring water tractor trailer accident) to freeze. The roadway does not freeze because the ground below causes it to freeze; the ground is typically a source of relative warmth in the winter. Bridges and overpasses have another freezing option, though—the *air* below.

When the air temperature is below freezing, roadway temperatures are often above freezing because the ground holds warmth longer than does air—the roadway is being warmed from below like the bathroom floor in a fine hotel. The surfaces of bridges and overpasses, though, quickly cool to the air temperature because they lack the insulation of the ground; they're exposed to air from above and below, rapidly matching the temperature of the air.

Normally, the exposure to the air below is not a serious problem (as in July); however, when snow is falling and melting on the roadway, ice might be collecting on bridges and overpasses. And when a dense fog forms with temperatures near freezing, bridges and overpasses might become icy.

Dense Fog All Day

When Mother Nature wants to slow us lowly human beings down, she often waves her magic weather wand and covers a region in dense fog. Air traffic stops, road traffic slows to a crawl, and even spying on our neighbors becomes impossible. Fog typically happens in a tranquil weather pattern (see "In a Fog" in Chapter 1), so we don't always expect it. She's usually kind enough to limit this disruption to the morning hours in the eastern part of the country, but in the West, dense fog can last all day long and persist for days.

Valley fog is common anywhere mountains exist. The fog forms when the calm wind of a high pressure system allows the night air to form into a cloud that settles in the valley. As long as there is enough moisture—from the air, from plants, or from moisture left from recent rain or snow—and the night is long enough to allow enough time for the cooling and condensation to occur (more likely in fall and winter than in summer), fog will form in a valley. Think of the fog as sand settling in the bottom of a bucket of water when it's no longer being stirred.

This fog typically dissipates the next morning, as sunshine begins to drive the normal breezes of the day, similar to how the sand would rise from the bottom of the bucket if a strong spigot were adding water to the bucket, mixing the water. The valleys in the western part of the country are much deeper than the valleys in the East because the mountains, such as the Rockies and Sierra, are much higher. As a result, the usual atmospheric mixing that breaks up the fog doesn't always dissipate the fog in the deep valley. Think of our bucket of water again: If sand sank to the bottom of a very deep bucket and the spigot adding water to the top of the bucket wasn't strong enough to create a current at the bottom, the sand would remain undisturbed.

That's what often happens in the valleys of the West, including the Central Valley of California, the valleys of interior Washington and Oregon, the Idaho valleys, and even some of the valleys in Nevada and Utah. Once dense fog forms at night, it might last through the next day, so travelers need to deal with dense fog on the morning commute *and* the afternoon com-

Warm in the Mountains,
Cold in the Valleys

The result of the valley fog is often the reverse of the normal temperature scheme. Normally, we expect it to be colder in the mountains than in the valleys, but when the valleys fill with sun-blocking fog and low clouds, it can be cold, damp, and dreary. In the mountains, above the fog, the days might be filled with sunshine and mild temperatures. For example, in California, valley locations such as Redding, Fresno, and Bakersfield might have high temperatures in the 30s, while mountain locations might have bright sunshine and high temperatures in the 50s.

mute. As long as the weather pattern doesn't change, fog or low clouds will persist, sometimes for a week or two. If the fog does begin to break late in the day, then it's likely to return in full force again at night, repeating the same foggy pattern. If it doesn't, then while the fog will not be as dense as it was to start, low clouds will linger.

Don't Raisin on My Parade

Raisins are dehydrated grapes, of course, but not every grape-growing location produces raisins. California has a favorable climate for both, which is why California is sometimes called the raisin capital of the world.

Grapes are often turned into raisins in the field, not in buildings with massive food dehydrators, so for good raisin production, a season of good grape production needs to be followed by a period of dry weather to turn the grapes into dried up, wrinkly grapes. (Since that's not a good name for a good product, someone came up with the word *raisins.*) Most of the grapes grown close to the coast in California are used for wine, and most of the grapes grown in the Central Valley are used for raisins.

Summers are notoriously dry in California, so by fall, most farmers are looking for rain—to get a head start on the winter rain—but raisin growers are one notable exception. Generally from late August through the middle of October, they want as many sunny, warm days (with low humidity) as they can get.

The biggest risk of rain from California during this time of year is from the south—the remnants of an eastern Pacific tropical storm or from moisture from the Desert Southwest—but early winter-like storms can bring rain to California from the north.

Of course, for a farmer with a year's worth of crop sitting vulnerably under the sky, it doesn't matter where the rain might come from—just so it doesn't come at all.

Free Sample: $3

Free wine samples are the perfect reason to tour a vineyard, so imagine my surprise when touring an Ontario Vineyard in finding a free sample that cost $3. Fine, it obviously wasn't a *free* sample, but of all the wines at the vineyard, the sample of ice wine was the only one so expensive to produce that the vineyard felt the need to charge for a tasting.

All wine is weather dependent of course, but ice wine has an unusual relationship with the weather—the wine is made of grapes left on the vine until *after* they freeze. Below-freezing temperatures at a vineyard are generally as popular as the flu outbreak during a hospital workers' convention, but wine made from these frozen grapes is intensely sweet, with a high alcoholic content because the process of freezing on the vine removes excess water and concentrates the sugar. I don't know whether this was an accidental discovery, but ice wine can be produced only in the colder climates of the world (of course). In addition to Ontario, ice wine often comes from Germany, but U.S. ice wine is produced in upstate New York.

It's worth noting that the expense of the ice wine is not directly related to the weather; it's related to grape yield. Leaving the grapes on the vine for a longer period of time decreases the number of grapes left by wine-making time: More grapes are lost to birds or other animals, and more grapes either rot on the vine or fall to the ground.

Good Weather Brewing

Brewing beer at home has become a popular hobby in the past couple of decades—even though it was probably more popular during Prohibition. True brewers may even want to grow their own hops (the flower of a hop vine), which is used almost exclusively in beer production.

The United States is a leader in worldwide hops production because much of the growing needs for hops can be met in main agricultural areas. There must be 120 frost-free days of growth, along with warm temperatures and ample sunshine. The tricky part is that hops need a fair amount of rain but also require well-drained soil. The Pacific Northwest is an excellent meteorological location for hops growing. The warmth, sunshine, and soil type are provided by Mother Nature, and the small amount of natural rain is supplemented (and controlled) by irrigation.

The home grower, who doesn't have the luxury of a rain-out-of-a-hose system, has to hope that the summer isn't too wet; otherwise, pests and disease will limit his ability to grow hops—and be the most popular person on the block.

How Hot *Was* It?

The punch line might not live up to the late Johnny Carson's standards, but "It was so hot that the roads buckled!"

While we might expect that temperatures would have to be over 100°F—perhaps even over 105°F—for roads to buckle, it can happen with temperatures in the 90s. When an early-season heat wave (late May) occurred in Minnesota and Wisconsin in 2006, a large section of I-94 buckled, resulting in road closures. The temperature hit 97°F in Minneapolis, Minnesota, on May 27, with back-to-back record highs of 95°F in Eau Claire, Wisconsin, on May 28 and May 29.

A contrast in temperatures contributes to bumpy weather, and this contrast was important in the bumpy road as well. The ground warms and cools much more slowly than does air, so engineers believe that the relative coolness of the spring ground

Cold Weather and Potholes

Cold weather is blamed for potholes (holes in roadways) across the colder parts of the country, but it's not technically the cold that causes the problem; it's the combination of moisture and temperatures alternating between below and above freezing.

Once a roadway develops a small crack, it allows a space for water (from rain or melted snow) to collect. When temperatures drop below freezing, the water inside of the crack freezes, and since water expands when it freezes, it enlarges the crack. When it warms again (a black surface of the road will rise above freezing, especially on a sunny day, long before the official air temperature rises above 32°F), the ice melts, leaving behind a larger crack. The larger crack can then accept a greater amount of water, which freezes and expands.

This vicious cycle is completed with a rash of Men at Work signs in the warmer months.

contrasting with the extreme heat of the spring air is what led to the buckled roadway.

No Sour Grapes in California

If a meteorologist were a sommelier, then the state of California would be the perfect weather cellar, with a microclimate for every weather preference. These microclimates assist the true sommeliers—the ones who deal in wines—by providing the types of climates necessary for the production of many types of wine grapes.

All states produce at least a little wine, but there is only one California. Sunny, dry weather is a prerequisite for grape production, and except for areas along the coast, this describes summer in California. The sunshine ensures proper sugar content, and the winter rain sets the stage for the grapes to grow in the summer dry weather.

Warm days and cool nights are best for grapes, and the marine influence for the valleys near the coast in northern and central California keep the weather from being too hot during the day, and the dry air allows for surprisingly cool nights (think desert). Heat that is too intense means that the grapes mature too quickly, before the flavor peaks; however, there are enough heat-resistant types of grapes to allow for grapes to grow in the Central Valley (see "Don't Raisin on My Parade" earlier in this chapter).

Sunny, warm, and dry weather can be found in many parts of the country during the summer; in fact, there are other great wine-growing regions in the United States. However, few regions match California for the consistency of this type of weather year after year, especially one with a cooling ocean close enough to bring moderation when needed.

No Wine-ing About a Drought in the East

Droughts can be some of the most destructive and costly of all weather disasters, so there is nothing humorous about a drought—and all parts of the country are susceptible. The perception of a drought in the East and West can be vastly different, however. In the East, 3 weeks of dry weather might be considered a drought, while in the West, 3 months of dry weather is considered the start of summer.

Western farmers are more prepared for dry weather because of irrigation or because they grow crops that require less water, such as grapes (see "No Sour Grapes in California" earlier in this chapter), but farmers in the eastern part of the country have less protection against drought. Eastern grape growers, however, are one group of farmers who are not likely to complain about a drought.

Dry weather is always favorable for grape growth, and the heat that I talked about being a potential problem in the West (because it makes the grapes mature too quickly) can be an advantage in the shorter growing season in the East.

Please Pass the Salt—or the Sand, Gravel, or Ash

Which came first: the road or the substances to clear the road from snow and ice?

Salts and other chemicals are often used to melt an accumulation of snow and ice, working by lowering the freezing point of water. Depending on the substance used, these do not always work when it's extremely cold or after the mixture has been diluted by the water of melting snow and ice. When a road crew knows snow is going to fall, roads are sometimes pretreated with a chemical that will melt the snow as it falls.

When it's too cold for salt to work effectively (generally around 10°F), or when improving traction is the goal, sand or gravel is often used; the darker color of such materials absorbs energy from the sun (even when it's cloudy, this will work to some extent) to melt the snow and ice on the road. Ash, which is black, is also used for this purpose. The limitation of this approach is that if the road is not dry by the time the sun sets, it can refreeze at night (assuming that temperatures drop below 32°F).

Road crews have the difficult job of making the roadways safe enough for us to resume normal use during and after a storm, which typically includes driving while talking on the phone, eating dinner, and helping the kids pick out a good DVD.

Snowbound Express

"All aboard! Passengers heading westward on the Snowbound Express, your train is ready to depart from the station. Next stop—the top of the Sierra for six fright-filled days." That's what could have been said when the passenger train *City of San Francisco* departed Chicago (or any of its other destinations) in January 1952.

The train was relatively close to its destination as it climbed Donner Summit, high in the California Sierra, but it was stopped in its tracks (literally!) by a massive snowstorm on January 13. As often happens in the mountains of the West, the storm—or, more accurately, the series of storms—resulted in a nearly constant ravage of snow and wind. By January 19, when rescue crews were finally able to arrive from the west, 12 feet of snow had accumulated at Donner Summit, whipped into massive drifts by a wind that gusted to 100 mph.

The passengers and crew, 223 people in all, had to survive frozen water (the result of which was a lack of properly operating toilets and a lack of drinking water), a minimal food supply, and no heat—and we complain when we're delayed at the airport for 2 hours.

Sweet Freeze

Having temperatures near freezing in a citrus orchard is analogous to giving a 15-year-old boy charcoal, lighter fluid, and a pack of matches: While the result might be abundant, perfectly grilled food, it could also be an irreparable disaster.

A light freeze, which means low temperatures of 29°F to 31°F for a couple of hours, can result in sweeter citrus than if temperatures remained above freezing. The slightly below freezing temperatures result in a higher sugar concentration within the fruit. The danger, of course, is that if temperatures remain below freezing for too long or if the temperature drops to slightly lower levels, say 28°F or 27°F, the fruit will begin to freeze—and not in a 100% juice from concentrate sort of way.

The line between a freeze that sweetens the fruit and one that damages the fruit is extremely narrow, not only because it's difficult to forecast the temperature accurately within a degree or two for a given location but also because the temperature within the confines of an orchard can vary significantly.

When a meteorologist forecasts a low temperature (or even

a range of low temperatures), it represents the forecast temperature for a given location, such as the middle of town or the airport; it by no means represents the wide range of temperatures throughout the entire region, which—especially on a clear, calm night—can vary dramatically over surprisingly small distances.

We've all probably noticed stark temperature changes when going for a walk on a clear evening, when it was much chillier at the bottom of a slight hill or when walking through an area with a light breeze was noticeably warmer. Those same types of differences are more likely to occur in wide-open areas, which is where farmers tend to have orchards for some inexplicable reason. Temperatures can also vary dramatically (again, most dramatically on clear, calm nights) from near the ground to several feet above the ground. You won't notice that on your evening walk, but a difference in temperature from the ground to 6, 8, or even 12 feet above the ground makes a big difference for citrus farmers because the fruit resides at those minutely higher elevations.

Farmers have ways of trying to combat this colder weather, including setting controlled fires to warm the air, trees, and fruit; spraying water on the trees and fruit, since a thin layer of water (or, if cold enough, ice) provides insulation from the cold; and blowing large fans to keep the air mixed (which can prevent the coldest air, which is the heaviest air, from settling in the orchard). These methods work to only a certain degree, leaving fruit farmers to shiver through a night when the change of a degree or two could mean the difference between making money with sweet fruit or losing a significant portion of their crop.

Ultimate Battle—Washington State Versus New York State

While the battle of the thin-crust New York–style pizza versus the deep-dish Chicago pizza is more famous, the battle of the apple—New York State versus Washington State—is nearly as ferocious.

Most of us think in terms of drastically different climates when talking about the states of Washington and New York. We think of Washington as a state with one season, wet (see "It Always Rains in Seattle" in Chapter 2), and New York as having the traditional four seasons. The growing seasons, though, are not that much different; in fact, New York State is the one more likely to have too much rain during apple growing season. The best weather for apple growing is warm days and cool nights, with a moderate amount of rain. The high humidity in the East makes apples more susceptible to disease and pests and keeps the nights warmer, but many people believe that the richer soil results in a more flavorful apple.

Discussion about flavor—that's the part that sounds similar to the aforementioned pizza battle.

The Weather Is Everywhere

HEALTH AND HOME

THE WEATHER—OR AT LEAST weather-related concepts—is more than just understanding that hot air rises or how hurricanes form; it's also about understanding how the weather and changes in the weather affect our lives, homes, and health. When you think of the weather in those terms, it's everywhere.

The difference between the relative humidity inside of the house and outside of the house can affect our homes, health, and hobbies. How our bodies react to the weather affects our health and safety, and understanding how the weather affects the world around us can affect the way we live our lives.

Can You Turn Down the Heat, Grandma?

Not only do we want to ask Grandma to turn down the heat in the winter, but we usually want to ask her to turn up the air conditioner in the summer. But we humans don't just develop an inability to cope with cold when we get older; we also develop an inability to cope with the heat.

A normal, healthy body adjusts to cold weather by dilating blood vessels, which allows more warming blood to flow, and we adjust to warm weather by increased sweating, which allows the cooling process of evaporation. Older people don't do either well; it's as if their internal thermostat needs to be replaced.

That's why it's important to monitor the health and safety of our older family members and neighbors during times of extreme heat and extreme cold.

Dehydration

Once upon a time, a young meteorology student who liked to play tennis wasn't very aware of the dangerous effects of dehydration on his body. This young man once played tennis for hours in the middle of a hot summer day without drinking any water. (The school was locked, so there was no access to the water fountain he normally depended on.) When I—I mean he—started to get severe chills in the middle of the fifth set, he knew that he'd done something very stupid. He had given himself heat exhaustion, and had he not heeded those warning signs (chills, nausea, fatigue, and weakness), he may have developed heat stroke, which can be life threatening.

Heat exhaustion and heat stroke are extreme forms of dehydration, but dehydration can cause myriad health problems, ranging from simple headaches to muscle cramps to irritability to the more severe cases described above. Of course, everyone should know the dangers of exercising without proper hydration (hey, I was young and stupid!), but just the weather by itself can lead to dehydration problems.

Sweating is our body's cooling mechanism, which we depend on most when it's hot. We might assume that we lose more water when it's hot and humid than when it's hot and dry, since we sweat more, but that's not necessarily the case. We might feel more comfortable when it's not humid, but we're certainly in no less danger of becoming dehydrated.

In the winter, when it's cold, we might think that we lose too little moisture to become dehydrated, but that's not the case. When the air is dry (and it's sometimes extremely dry in the winter), we lose more moisture than when the air is moist, so dehydration is still a concern. The weather most certainly affects our hydration levels, so take that into consideration whenever the air is hot or exceptionally dry.

In other words: Do as I say, not as I did when I was younger.

Frozen Pipes

While record-breaking cold outbreaks are a serious problem for many (such as farmers), they're often a boon for one occupation:

the plumber. Intense cold often results in a rash of frozen and bursting water pipes in homes and businesses.

The pipes themselves are not of poor quality, and the cold air in these regions is not magically more destructive, of course. Pipes in many of the buildings (especially the older ones), though, are not properly protected against cold weather because extreme cold is rare. The pipes might run through an uninsulated attic or basement wall or might be located outside of the house rather than under the ground.

Regardless of the lack of these normal safeguards, pipes inside of a home do not typically begin to freeze until air temperatures drop to 20°F or lower. Water, of course, will freeze at 32°F, but the temperature even in an uninsulated building will be 37°F or 38°F when it's 32°F outside. In addition, just as a lake will not freeze solid at 32°F, all of the water inside of a pipe will

Busy Nights for Southern Plumbers
(All-Time Record Lows for Southern Cities)

Palm Springs, California	January 22, 1937	19°F
Brownsville, Texas	December 23, 1989	16°F
Orlando, Florida	January 21, 1985	19°F
New Orleans, Louisiana	December 23, 1989	11°F
Savannah, Georgia	January 21, 1985	3°F
San Jose, California	December 22, 1990	19°F
Phoenix, Arizona	January 7, 1913	16°F

not freeze at 32°F. It will, however, freeze solid much more quickly than a lake would.

When the temperature drops below 20°F outside, though, it's best to allow water to dribble out of faucets attached to uninsulated pipes. Moving water doesn't freeze as quickly as standing water, and even the ice that forms is less likely to develop pipe-bursting pressure because some water is moving.

Hypothermia

Hypothermia is a lowering of the body temperature; it is not an exposure to extremely cold weather. Of course, running around in shorts, a T-shirt, and sandals when it's 2°F outside might be a certain way to become hypothermic (which can kill a person, by the way), but it can happen when exposed to other weather conditions as well.

The normal body temperature is 98.6°F, so in theory, hypothermia could begin to happen at any temperature below that point. In fact, if a person is in extremely poor health, has an exceedingly poor level of fitness, or has serious circulatory problems, then a reduction in body temperature can occur when it's mild outside, say 70°F.

Those are rare cases, but hypothermia can be a serious problem in surprisingly mild temperatures when an individual is wet, because water removes heat from the body 25 times faster than when a person is dry. Therefore, hypothermia can be a concern for anyone caught in the rain without water-repellant clothing at temperatures in the 40s, 50s, or even 60s. In addition, heat is lost not only to the air but also to other surfaces with which a person comes in contact, such as rocks or snow-covered ground.

Ice at 34°F

It might be tempting, but don't put your ice scraper/snow brush back in the garage just because the temperature is 34°F and rain is on the way—and don't even think about changing from the rubber-soled shoes back to the stiletto heels!

Not only are icy car windows still possible with air temperatures above 32°F, but even sidewalks and roadways can become slippery at temperatures of 33°F, 34°F, or perhaps 35°F when rain begins to fall after a period of cold weather. After being exposed to subfreezing temperatures for an extended period of time, it will take a while for such surfaces to warm enough to match the air temperature—think of sitting on a metal bench in a park in the morning after a cold night! Before sidewalks and roads warm, rain will freeze upon contact with the cold surfaces. Ice at above-freezing temperatures is more likely to happen when rainfall is light because a downpour would quickly warm the surface to temperatures above freezing (the temperature of each water droplet is warmer than 32°F).

It's Not the Hair; It's the Humidity

I hate high humidity as much as the next person, but I sometimes feel sorry for this moist monster because it gets blamed for everything, including our most common cosmetic disaster: bad hair. I can't be too sympathetic, though, since high humidity levels do, indeed, affect our hair.

Squeezed into each tiny strand of hair resides two proteins, and both of these expand when exposed to the additional moisture of a high humidity day. They don't expand in a uniform manner, though; it varies from person to person and from strand to strand. When one part of the hair expands more than another, the hair bends or stretches, falling differently from the way it normally does. Straight or curly hair often becomes frizzy, but gently curly or wavy hair might become flat.

Talk about a difficult forecast; I'm glad that I didn't choose a career as a cosmetologist instead of a meteorologist.

Home-idity: Home Relative Humidity

Dew point temperature and relative humidity (see "Relative Humidity/Dew Point" in Chapter 1) are the two ways to gauge the amount of moisture in the air. Of the two, the dew point temperature is much more important to a meteorologist, and the relative humidity is much more important to a home owner.

For a home owner, though, the relative humidity reading outside, reported on the evening news, is not as important as the relative humidity inside the house. Since relative humidity depends on air temperature, when we warm or cool our homes, the relative humidity levels inside can be significantly different from those outside.

In the summer, the cooler air inside of our homes has a

Paint Time

For indoor and outdoor painting interests, temperature, wind, and humidity are the three weather factors that need to be considered—other than the obvious, such as that outdoor painting can't be done while it's raining.

The temperature of the air and the surface being painted should not be below 50°F or above 90°F to ensure that paint adheres properly, which presumably is rarely a consideration with indoor painting. The wind is also not a consideration for indoor painting, but it is outside. Not only can a strong wind blow debris into the wet paint, but it can also cause the paint to dry too quickly, resulting in an inconsistent finish.

That leaves relative humidity, which is a concern for both indoor and outdoor painting. When it's too humid for paint to dry effectively, condensation can form on the painting surface before it dries, leaving streaks, inconsistent color, and a tacky surface. Rainy days don't automatically disqualify indoor painting. When it's cool and rainy outside, the inside air can perhaps be warmed enough to lower the humidity to a level that is acceptable for painting.

Buy a hygrometer and follow the instructions on the paint can.

HYGROMETERS

Relative humidity is determined using a calculation based on air temperature and humidity, but home owners don't need to fuss with that. Buy a hygrometer, which measures the relative humidity. In fact, a hygrometer is often included in the purchase of a humidifier.

much higher relative humidity reading than the air outside. For instance, if the air temperature outside is 95°F with a dew point of 66°F, the relative humidity is 31%, but by cooling the air temperature inside to 72°F, the relative humidity becomes 70% to 75%. (The dew point would be a little lower because air-conditioning decreases the amount of moisture in the air, but it will not be significantly lower than the air outside.)

To prevent problems associated with high relative humidity, such as mold and mildew, condensation on toilet tanks (see "Peak Toilet Tank Condensation Season" later in this chapter), and poor air quality for those with asthma or other respiratory illnesses, many of us need dehumidifiers in the summer.

The converse is true in the winter. If the air temperature outside is 35°F with a dew point temperature of 29°F, then the relative humidity is 71%, but with the air warmed to 68°F inside a building, the relative humidity is 15% to 20%. (The dew point will be approximately the same, but it might be a little higher than outside because of moisture added from respiration, toilet tanks, etc.) That low relative humidity results in certain problems inside of a house, including incredibly dry skin, virility of certain germs, static cling, and squeaking floors and doors. These are all cries for a humidifier.

My Knee Says It's Going to Rain

We all know people who claim that they can forecast impending precipitation based on an achy joint, such as a knee. We're also probably related to at least one person who thinks he or she is a walking barometer and forecasts the weather based not only on an achy knee but also on a cranky ankle, swollen knuckles, and a cantankerous elbow or two—and tells us all about all of it at every family gathering where someone makes the mistake of asking the seemingly simple question, "How are you?"

When the outside air pressure falls, as it does when a storm is approaching, healthy tissue expands; however, scar tissue does not expand as much, and this different rate of expansion places stress on joints, a medical explanation meaning joint pain.

In other words, friends and relatives are not complaining to hear themselves speak—well, not all of them.

News Flash: Cold Air Doesn't Make You Sick

Whether it's a television show, a novel, or a Hollywood movie, we've seen this tired scene hundreds of times: Someone who was forced to be outside in the cold for an extended period of time comes inside and sneezes once, and in an instant, he has a cold or the flu. If I were writing a book on health myths, the myth that cold air causes sickness would be in Chapter 1, on page 1.

Cold and flu are much more common in the winter than in other times of the year, so jumping to the conclusion that the cold weather *causes* the sickness is logical, but it's simply not true. We spend more time together in buildings without access to fresh air (windows are closed) during the winter, which results in an easy passing of germs from person to person. In addition, when it's cold outside, our tolerance to illness is lower because much of our body's energy is spent on keeping warm, not fighting illnesses. In other words, your mother was right about wearing the winter coat, hat, and gloves.

Finally, dry air is more conducive to certain types of bacteria and viruses, and air, especially inside of a house (see "Home-idity: Home Relative Humidity" earlier in this chapter), is much drier during the winter than during the summer.

Cold air contributes to conditions likely to spread viruses, but it doesn't *cause* a person to become sick.

Historic Bedfellows

One of our Founding Fathers, Benjamin Franklin, believed not only that cold air did not cause illnesses but that it was good for you. David McCullough, in the book *John Adams*, recounts a discussion between Franklin and the soon-to-be second president of the United States on a night that they were forced to share a small room while traveling:

"I answered that I was afraid of the evening air," Adams would write, recounting the memorable scene. "Dr. Franklin replied, 'The air within this chamber will soon be, and indeed is now, worse than that without doors. Come, open the window and come to bed, and I will convince you. I believe you are not acquainted with my theory of colds.'" Adams assured Franklin he had read his theories; they did not match his own experience, Adams said, but he would be glad to hear them again.

So the two eminent bedfellows lay side by side in the dark, the window open, Franklin expounding, as Adams remembered, "upon air and cold and respiration and perspiration, with which I was so much amused that I soon fell asleep."

Peak Toilet Tank Condensation Season

Meteorologists often talk about the peak of tornado season being May and June, the peak of hurricane season being late August through September, and the peak of winter being January and February. Why is it that no one talks about the peak of toilet tank condensation season (except for me, of course)?

When the first warm, humid air mass of the season arrives, often in May or early June, the contrast of the cold water in the tank with the warm, humid air results in condensation on the outside of the toilet tank. Water dripping from the side of the tank usually comes on quickly, and before you know it, you have an old towel lying on the floor for a couple of months.

While it will be humid much of the time in the summer across much of the country, the condensation problem is typically more serious through early summer, when the water in the tank is most likely to be at its coolest. I say "most likely" because tank water temperature varies based on the local water source. If it's a local reservoir, the water will be a little warmer by later in the summer, reducing the intensity of the condensation. If the water source is underground water, then condensation problems will remain largely the same all year, since underground water sources remain close to the same temperature all summer.

Better solutions than the old-towel-on-the-floor trick are available, including an insulating tank liner, a tray to catch the condensation, or a new home in Phoenix.

Shocking Development

As a child, one of my favorite pastimes was dragging my feet as I walked across the carpet to build up enough static electricity to shock an unsuspecting family member with my electric touch. I guess I didn't have enough hobbies.

When two dissimilar surfaces touch, there is a weak electrical exchange. This typically goes unnoticed unless one of the substances has a resistance to electrical flow, resulting in an accumulation of the electrical charge that is then released with a shock, such as when a bratty child touches someone else after dragging his feet on the carpet.

Static electricity is much more likely to occur when the air is dry, meaning during the winter for much of the United States.

Static Electricity Lightning in Dust Storms

While lightning is like static electricity on a much larger scale, there is a lesser-known pseudo-weather phenomenon that is much more like standard, household static electricity—static electricity lightning strikes that occur during dust storms.

During the Dust Bowl years, dirt particles rubbing together in the desert-like dry conditions of a dust storm resulted in an extreme accumulation of static electricity, which was sometimes released in the form of a miniature lightning bolt. The bolt, typically not strong enough to kill, was an additional hazard (such as a fire hazard) in the already unbelievably harsh conditions.

When the air is moist, a greater number of water molecules are present, and water, being a better conductor of electricity than air, allows the electricity to flow through the surface rather than accumulate. As a result, children can drag their feet all day long when it's humid and not cause any shocks, which is why they don't do it. I should know.

Skin-Freezing Cold

Confusion about temperature and windchill is about as common as confusion about humidity and relative humidity (see "It's 100°F with 100% Humidity" in Chapter 2), and it's just as understandable.

The windchill is how cold it *feels* (an estimation of the temperature, in other words) when the air temperature and wind speed are taken into consideration, and it's a concern only to living, breathing beings. When we sweat, the cooling process of evaporation of the liquid into the air cools the body. The same concept applies when an icy wind hits us in the winter cold. Although we certainly aren't helping the process by sweating, the wind is still removing moisture and heat from our bodies, lowering our internal temperature, which our bodies are fighting so hard to maintain. (Shivering is the body's attempt to remain warm.) Thus, we get colder and colder when the wind blows on a cold day. This is calculated by a mathematical equation and is known as windchill.

When our bodies are exposed to a windchill temperature of −19°F, our exposed skin (not the skin protected by layers of clothes, a hat, or earmuffs) will freeze in about 30 minutes. A windchill of −19°F can be achieved by an air temperature of −19°F and no wind, by an air temperature of 5°F and a wind of 30 mph, or by a temperature of 10°F and a wind of 60 mph.

Bright sunshine can increase the windchill by as much as 10°F to 18°F, but when it feels like −19°F outside, I won't be in it long enough to test that theory!

Sun from All Angles

I'm not saying that we're overly cautious about sun exposure at my house, but we use a sunscreen with SPF 100; when we open the bottle, clouds automatically appear.

Windchill Does Not Affect Cars, Water Pipes, and Bananas

The windchill does not affect inanimate objects, such as cars (or antifreeze, oil, or transmission fluid), water pipes, and fruit, since they don't have an internal air temperature that can be lowered. Inanimate objects will get as cold as the air temperature; they will not "feel" colder when it's windy.

When it's windy, the heat that these inanimate objects possess will be removed more quickly, so they will chill to the air temperature quickly; however, they will never get colder than the air temperature. In other words, pipes will not freeze when the windchill is 20°F (see "Frozen Pipes" earlier in this chapter); they will freeze only when the air temperature is 20°F.

The danger of sun exposure does not end with the end of the summer season; in fact, some of the worst sunburns can be had while enjoying snow-related outdoor activities in the winter. Many of these activities are done in the mountains, of course, where the air is naturally thinner, making sunburns more likely no matter the season. More important, skiers are much more likely to spend an extended period of time outside compared to the rest of us. Besides, with snow on the ground and sun in the sky, the sun is hitting skiers from all angles—not only from above but from below, since the sunlight is being reflected upward from the snow-covered ground.

See you on the slopes—that is, if we can get a sunscreen with SPF 125.

Water the Garden *After* It Rains

If a light or moderate amount of rain (less than half an inch) falls after a prolonged stretch of dry weather during the growing sea-

Vaseline to the Rescue

Dressing in layers helps keep us warm because it provides pockets of air insulation, but another way to keep warm in the winter is to apply a thin layer of Vaseline to the skin. This also provides a layer of insulation, which keeps the warmth of the body in and the cold of the air out. It's also one of the best hypoallergenic moisturizers available.

son, then my meteorological recommendation is that you immediately water your garden. While that might not seem to make sense, plants benefit best from a good soaking—generally 1 inch or more of rain. Not only does this ensure enough water for the plants, but it conserves water (you don't have to provide the full amount of water needed, as you would when the ground is dry) and promotes deep root growth, which is a natural protection against summer dry spells.

If you're certain that only a small amount of rain is going to fall (and there's not going to be any lightning), you could water the garden *while* it's raining. Not only will it have the same effect on the garden, but it will give the neighbors something to talk about.

We're All Snot-Nosed Brats When It's Cold Outside

When it's cold outside, the small veins inside of our noses dilate in order to bring more blood to the nose in an attempt to warm

Water Plants in the Morning or at Night

Generally speaking, it's best to water plants either at night or in the morning. Water evaporates more during the hottest part of the day, so morning and evening allow for the most effective use of water. In addition, sensitive plants may burn if watered while the sun is out during a hot day, since before the water evaporates, the droplets will act as miniature magnifying glasses, intensifying the sun's rays.

the air before it goes to our lungs. A side effect of this dilation is that the nose produces more mucus, with the result being that we can never carry enough tissues when it's cold.

I don't know about you, but I'd rather breathe in the unmoderated and cold air.

Resources

American Lung Association: www.lungusa.org
American Public Media: http://americanpublicmedia
.publicradio.org
The Beauty Brains: http://thebeautybrains.com
BrewingTechniques: http://brewingtechniques.com
Brigham Young University: www.byu.edu
Centers for Disease Control and Prevention:
www.cdc.gov
Chicago Tribune: www.chicagotribune.com
Exide Technologies: www.exide.com
KidsHealth: http://kidshealth.org
Los Angeles Times: www.latimes.com
**National Centers for Environmental Prediction,
Hydrometeorological Prediction Center:** www.hpc.ncep
.noaa.gov
National Hurricane Center: www.nhc.noaa.gov
**National Oceanic and Atmospheric Administration
(NOAA):** www.noaa.gov
National Weather Service: www.nws.noaa.gov
NOAA Central Library, U.S. Daily Weather Maps Project:
http://docs.lib.noaa.gov/rescue/dwm/data_rescue_daily_
weather_maps.html
Natural Handyman: www.naturalhandyman.com
Pennsylvania Highways: www.pahighways.com
Pep Boys: www.pepboys.com

Princeton University: www.princeton.edu
Storm Prediction Center: www.spc.noaa.gov
University of Chicago: www.uchicago.edu
USA Today: www.usatoday.com
U.S. Environmental Protection Agency: www.epa.gov

Index